116
Structure and Bonding

Series Editor: D. M. P. Mingos

Structure and Bonding

Recently Published and Forthcoming Volumes

Intermolecular Forces and Clusters II

Volume Editor: D. J. Wales

With contributions by
R. A. Christie · G. E. Ewing · B. Jeziorski · K. D. Jordan
K. Patkowski · K. Szalewicz · S. S. Xantheas

 Springer

The series *Structure and Bonding* publishes critical reviews on topics of research concerned with chemical structure and bonding. The scope of the series spans the entire Periodic Table. It focuses attention on new and developing areas of modern structural and theoretical chemistry such as nanostructures, molecular electronics, designed molecular solids, surfaces, metal clusters and supramolecular structures. Physical and spectroscopic techniques used to determine, examine and model structures fall within the purview of Structure and Bonding to the extent that the focus is on the scientific results obtained and not on specialist information concerning the techniques themselves. Issues associated with the development of bonding models and generalizations that illuminate the reactivity pathways and rates of chemical processes are also relevant.

As a rule, contributions are specially commissioned. The editors and publishers will, however, always be pleased to receive suggestions and supplementary information. Papers are accepted for *Structure and Bonding* in English.

In references *Structure and Bonding* is abbreviated *Struc Bond* and is cited as a journal.

Springer WWW home page: springer.com
Visit the SAB content at springerlink.com/

QD 461
.592
VOL·116
2005

Library of Congress Control Number: 2005930054

ISSN 0081-5993
ISBN-10 3-540-28191-6 Springer Berlin Heidelberg New York
ISBN-13 978-3-540-28191-7 Springer Berlin Heidelberg New York
DOI 10.1007/b100423

Springer is a part of Springer Science+Business Media

springer.com

© Springer-Verlag Berlin Heidelberg 2005
Printed in Germany

The use of registered names, trademarks, etc. in this publication does not imply, even in the absence of a specific statement, that such names are exempt from the relevant protective laws and regulations and therefore free for general use.

Cover design: *Design & Production* GmbH, Heidelberg
Typesetting and Production: LE-TEX Jelonek, Schmidt & Vöckler GbR, Leipzig

Printed on acid-free paper   02/3141 YL – 5 4 3 2 1 0

Structure and Bonding
Also Available Electronically

For all customers who have a standing order to Structure and Bonding, we offer the electronic version via SpringerLink free of charge. Please contact your librarian who can receive a password or free access to the full articles by registering at:

springerlink.com

If you do not have a subscription, you can still view the tables of contents of the volumes and the abstract of each article by going to the SpringerLink Homepage, clicking on "Browse by Online Libraries", then "Chemical Sciences", and finally choose Structure and Bonding.

You will find information about the

– Editorial Board
– Aims and Scope
– Instructions for Authors
– Sample Contribution

at springer.com using the search function.

Anthony Stone

Intermolecular Forces and Clusters:
Contributions in Honour of Anthony Stone

David J. Wales

Department of Chemistry, University of Cambridge, Cambridge CB2 1EW, UK
dw34@cam.ac.uk

It is a great pleasure to introduce this collection of papers in honour of Anthony Stone. The title of the collection reflects Anthony's ground-breaking contributions to our understanding of intermolecular forces [1], and how this work has been exploited to describe the structure and dynamics of molecular clusters. I personally feel indebted to Anthony for all the insight he has given me over my career, starting with undergraduate lectures that set the highest standard for clarity and accuracy. I know that the contributors who were fortunate enough to be guided by Anthony for either PhD or post-doctoral projects share these feelings, and several of them are represented by chapters in this collection. The two volumes of *Structure and Bonding* in this series are completed by papers from other researchers who have been influenced or helped by Anthony in their endeavours. In organising this collection I found that similar unsolicited testaments to Anthony appeared from many of the authors. For example, his depth of understanding, patience in discussions, and quiet good humour were frequently mentioned. His service to the community, in developing and maintaining programs such as ORIENT, and in providing an outstanding textbook [1], has undoubtedly contributed to his far-reaching influence on the field of intermolecular forces.

Some of the contributions address the calculation of intermolecular forces at a fundamental level, while the majority are concerned with applications, ranging from water clusters, through surfaces, to crystal structures. Szalewicz, Patkowski and Jeziorski provide a timely review of how perturbation theory can be used to address intermolecular forces in a systematic way. In particular, they describe a new version of symmetry-adapted perturbation theory, which is based on a density functional theory description of the monomers. The interpretation of bonding patterns for both intra- and intermolecular interactions is addressed in Popelier's review, which focuses on quantum chemical topology. He suggests a novel perspective for treating several of the most important contributions to intermolecular forces, and explains how these ideas are related to quantum delocalization.

Water clusters continue to be a particularly active area of research for both theory and experiment. The chapters authored by Xantheas and by Christie and Jordan both address the structure and bonding in water clusters using *ab initio* electronic structure theory. Christie and Jordan show how an *n*-body decomposition of the binding energy can be used to perform MP2-level calculations for clusters with up to fifty water molecules. Xantheas also considers an *n*-body decomposition, and analyses different approaches to the systematic development of empirical intermolecular potentials based upon *ab initio* data. Millot's review extends some of these ideas to molecular dynamics simulations using both empirical and quantum mechanical treatments of the electronic structure. He compares results obtained for hydrated halide ions within these different frameworks, and also considers dynamical simulations of silicon clusters.

Ewing also explores the theme of hydration in his chapter, this time in terms of adsorption of water on the surface of a sodium chloride crystal. Here we see how intermolecular forces determine the efficiency with which salt is dispensed from a shaker, and learn that the actual mechanism by which salt dissolves in water is still poorly understood. Tsuzuki's review concerns intermolecular forces that involve the microscopic surfaces of aromatic rings. These interactions are important for molecular recognition in biological systems, and play a major role in determining the crystal structure when aromatic rings are involved. Crystal structure prediction itself is addressed directly by Price and Price. They review the critical role of the intermolecular potential in providing useful predictions to discriminate between polymorphs, and highlight recent progress in this field. Here the distributed multipole analysis and distributed polarizabilies, pioneered by Anthony Stone [1, 4–10], play a crucial role in the development of accurate potentials. The application of such ideas to the structure of condensed matter provides a logical extension of earlier work for van der Waals complexes [11, 12].

Anthony Stone has also made influential contributions to fullerene research and to the bonding in inorganic clusters. The chapter by Soncini, Fowler and Jenneskens considers ring currents and aromaticity in fullerenes, using selection rules based on angular momentum theory to determine the aromatic and anti-aromatic terms. This approach ties in neatly with Anthony Stone's tensor surface harmonic (TSH) theory, which enables us to understand and predict electron counts of main group and transition metal clusters [13–16]. It also prompts me to conclude this overview with some of my own reminiscences. The TSH theory was largely worked out before I began my own PhD with Anthony in 1985. However, I remember that in my first term of research Anthony received a suggestion from Christopher Longuet-Higgins (his former PhD supervisor [17]) that it might be helpful to apply a symmetry-based approach to the π system of C_{60}. This idea proved to be very fruitful, and we quickly produced a descent-in-symmetry picture for C_{60}, based on the wavefunctions for a free particle-on-a-sphere. At the time the suggestion that C_{60} might be particularly stable as a truncated icosahedron [18] was unproven. Furthermore, the descent-in-symmetry

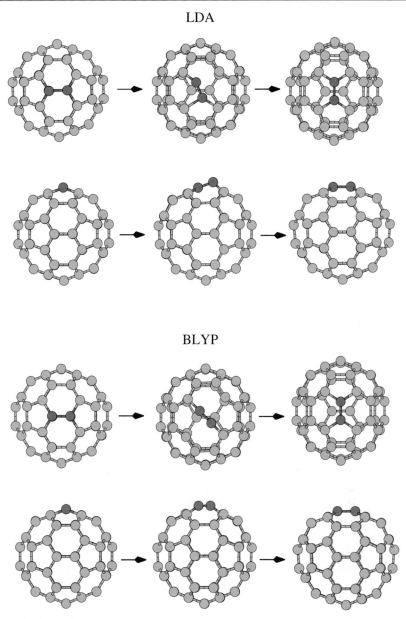

Fig. 1 Top and side views of the pathway connecting the lowest two fullerene isomers of
C_{60} calculated with a plane-wave density functional approach using the LDA and BLYP
functionals [2]. The middle panel is the transition state and the end panels are the two
minima in each case. The asymmetric path was calculated using the LDA functional
with a cutoff of 30 rydberg and the symmetric path corresponds to the BLYP functional
and a cutoff of 40 rydberg [2]. Reproduced with permission from D. J. Wales, *Energy
Landscapes*, Cambridge University Press, Cambridge (2003)

approach suggests that other reasonably spherical arrangements of carbon atoms should have similar π delocalisation energies. We therefore continued our study by considering a few alternative C_{60} fullerenes, and as an afterthought we decided to think about how such structures might interconvert [19]. The main result was the process illustrated in Fig. 1, which has since come to be known as the 'pyracylene' or 'Stone-Wales' (SW) rearrangement [20, 21].

The most recent calculations conducted to characterise the SW process have identified two possible transition states, both corresponding to concerted, single step mechanisms with similar barriers [2, 22] (Fig. 1). Furthermore, the same generic mechanism connects a whole hierarchy of C_{60} fullerenes [23, 24]. The energy of these structures increases systematically with the number of rearrangements, and when combined with the large barrier height produces the 'willow tree' disconnectivity graph shown in Fig. 2 [2, 25].

The urge to study rearrangements and global potential energy surfaces was therefore imprinted upon me at an impressionable age, and has influenced virtually all my research ever since. I am therefore led to reflect almost every

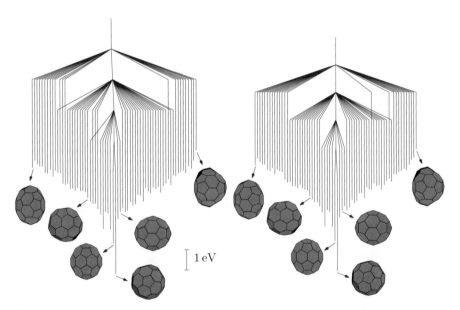

Fig. 2 Disconnectivity graphs for minima and transition states in the five lowest Stone-Wales stacks [3] of C_{60} calculated using a plane-wave implementation of density functional theory [2]. The graphs on the left and right correspond to the LDA and BLYP functionals, respectively. The vertical energy scales are the same and the energy zero has been shifted to buckminsterfullerene in both cases. The structures of six minima are indicated, including buckminsterfullerene and the next-lowest structure with C_{2v} symmetry. Reproduced with permission from D. J. Wales, *Energy Landscapes*, Cambridge University Press, Cambridge (2003)

day, on how lucky I was to have Anthony Stone as a mentor on this journey. Although it was not easy to persuade him that his achievements should be highlighted, I hope he will be happy with the following tributes from some of his friends.

June 28, 2005   David J. Wales

References

1. Stone AJ (1996) The Theory of Intermolecular Forces. Clarendon Press, Oxford
2. Kumeda Y, Wales DJ (2003) Chem Phys Lett 374:125
3. Austin SJ, Fowler PW, Manolopoulos DE, Zerbetto F (1995) Chem. Phys. Lett. 235:146
4. Stone AJ (1981) Chem Phys Lett 83:233
5. Stone AJ (1985) Mol Phys 56:1065
6. Stone AJ, Alderton M (1985) Mol Phys 56:1047
7. Price SL, Stone AJ (1987) J Chem Phys 86:2859
8. Stone AJ (1989) Chem Phys Lett 155:102
9. Stone AJ (1989) Chem Phys Lett 155:111
10. Williams GJ, Stone AJ (2003) J Chem Phys 119:4620
11. Buckingham AD, Fowler PW (1985) Can J Chem 63:2018
12. Buckingham AD, Fowler PW, Stone AJ (1986) Int Rev Phys Chem 5:107
13. Stone AJ (1980) Mol Phys 41:1339
14. Stone AJ, Alderton MJ (1982) Inorg Chem 21:2297
15. Stone AJ (1984) Polyhedron 3:1299
16. Mingos DMP, Wales DJ (1990) Introduction to Cluster Chemistry. Prentice-Hall, Englewood Cliffs
17. Mills I, Salem L, Stone A (2005) Mol Phys 103:141
18. Kroto HW, Heath JR, O'Brien SC, Curl RF, Smalley RE (1985) Nature 318:162
19. Stone AJ, Wales DJ (1986) Chem Phys Lett 128:501
20. Fowler PW, Manolopoulos DE (1992) Carbon 30:1235
21. Fowler PW, Manolopoulos DE, Ryan RP (1992) J Chem Soc, Chem Comm 408
22. Bettinger HF, Yakobson BI, Scuseria GE (2003) J Amer Chem Soc 125:5572
23. Liu X, Klein DJ, Seitz WA, Schmalz TG (1991) J Comp Chem 12:1265
24. Manolopoulos DE, May JC, Down SE (1991) Chem Phys Lett 181:105
25. Wales DJ (2003) Energy Landscapes. Cambridge University Press, Cambridge

Contents

Contents of Volume 115

Intermolecular Forces and Clusters I

Volume Editor: D. J. Wales
ISBN: 3-540-28194-0

Struc Bond (2005) 116: 1–25
DOI 10.1007/430_012
© Springer-Verlag Berlin Heidelberg 2005
Published online: 1 November 2005

H₂O on NaCl: From Single Molecule, to Clusters, to Monolayer, to Thin Film, to Deliquescence

George E. Ewing

Department of Chemistry, Indiana University, Bloomington, IN 47405, USA
ewingg@indiana.edu

1
Overview

In selecting a box of table salt from a grocer's shelf, the buyer is informed in big letters, that the product either does or does not supply iodide, a necessary nutrient. When the salt does contain iodide, a trace of KI has usually been added. However, even if the product has not been fixed up with a bit of iodide, there are almost always additives as listed on the side of the salt box in fine print, typically calcium silicate, yellow prussicate of soda, or sodium silicoaluminate, that are said to be anticaking agents. These additives as coatings to the grains have been applied to foil the consequences of adsorbed water that is present on the salt under ambient conditions. This adlayer water can lead the grains to stick together, making their dispensing from a salt shaker difficult. If the humidity is high, the grains without the anticaking agent undergo deliquescence as beads of liquid (brine) form during dissolution of the salt. A shaker of salt under these conditions becomes a gloppy mess.

Adsorption of water, leading to a thin film layer, is not peculiar to the NaCl surface, but is ubiquitous. Thin film water can coat metals, semiconductors, insulators, even ice under ambient conditions [1–6], and the thin film can have a profound effect on the physical and chemical properties of the sub-strates it covers.

We have selected water on salt as the subject of this review for a number of reasons. In general it serves as a model system for the study and understanding of water on a surface. The particular pairing of solid salt with water is relevant in a culinary context, as we have illustrated in the introductory paragraph. However, on a much larger scale, there is the industrial production of NaCl annually of 2×10^8 metric tons [7] that needs to be handled and transported in an often humid environment. Sea salt is produced naturally in even greater amounts from the oceans of the earth, and is generated at an annual rate of total mass flux of 5×10^9 metric tons [8], thus exceeding industrial production by an order of magnitude. These sea salt particulates, with their accompanying adlayers of water, undergo an intricate chemistry in the troposphere with natural and anthropomorphic trace gases. However, the primary focus of this review will be from the perspective of the chemistry and physics of the carefully prepared NaCl surface and its interactions with water. Many groups have participated in this science. A wide variety of the experimental studies of water on salt surfaces will be discussed in this review. Particularly important to the understanding of this intricate system have been the theoretical contributions of Dr. Anthony Stone, to whom this volume is dedicated, as well as other groups, as we shall see in the sections to follow.

We begin first with a discussion of the surface of NaCl before its exposure to water.

2
The NaCl Surface

Consider what the surface of salt looks like in the absence of water. The picture that one obtains inevitably depends upon how the salt is prepared. When poured from a box of table salt, the crystallites appear to the eye as a collection of remarkably uniform cubes 0.3 ± 0.1 mm on a side. When coated with a thin film of gold and examined by scanning electron microscopy (SEM) [9] the salt resembles that in Fig. 1.

Under this magnification, each crystal can now be distinguished from its neighbors by characteristic imperfections. A typical surface is in general flat, but the terraces are frequently marred by pits or bits of smaller crystals stuck to them. The terrace edges are rounded, dropping to lower ledges defined by perpendicular faces. Thus, table salt prepared by precipitation from water solution and dried [10] has surfaces that abound with imperfections and defects.

When salt is prepared conceptually, we can imagine beginning with an array of Na^+ and Cl^- ions arranged on a face-centered cubic lattice [11]. A slice of this ideal crystal can of course be directed along any plane, but generation of (001) faces is a natural choice, since this separation into two parts requires

Salt Crystals (Sodium Chloride) 135X @ 8"x8" size   Copyright © by David Scharf, 1977, 1978

Fig. 1 Scanning electron micrograph of table salt. From D. Scharf [9]

less energy than other possible divisions [12]. Ions at a surface experience an asymmetrical environment. In particular, strong electric fields arise from ions at and below the surface. Using the Lennard-Jones and Dent [13] analytical expression, the electric field only 0.24 nm above a Na$^+$ ion, where an H$_2$O molecule might fit, is easily calculated to be 8×10^9 Vm^{-1} and is orthogonal to the surface. Over the nearest Cl$^-$ ion, for the same vertical displacement, the electric field has an equal magnitude but has flipped its direction. Halfway between a Na$^+$ Cl$^-$ ion pair, but still 0.24 nm above the surface, the component of the electric field perpendicular to the (001) face drops to zero. Thus at distances where a small molecule like H$_2$O might be adsorbed to the surface, it will be enveloped in a tortuous electric field. While the electric field is large for distances close to the surface, at 1 nm it is 2 orders of magnitude smaller than for 0.24 nm. Calculations [14] show that a surface relaxation in response to this electric field causes rumpling. The Na^+ ions in the first layer are drawn in by 26 pm while the Cl^- ions are moved away from the surface by a smaller amount. A Cl^- ion at the surface develops a dipole of 1.5D (4.9×10^{-30} C m), and the less polarizable Na^+ ion a smaller value. Displacements of the ions from their original single crystal lattice positions and electronic distortions in the second and lower layers are more subtle. Surface defects, such as steps, edges and kinks, have been modeled [15]. Because cancellation of electric fields of nearby ions is less complete at defect sites, these sites are more susceptible to physical adsorption processes or even chemical activity.

The surface of small clusters, Na_nCl_m, containing ten or so ions, has also been described and explored theoretically where edges and vacancies control the properties of adsorbed molecules [16, 17]. Whereas Na^+ and Cl^- are isoelectronic with the rare gases Ne and Ar, respectively, and are anticipated to be inert, defects consisting of neutrals, Na and Cl, will exhibit chemical reactivity. Vacancies containing electrons, F centers [11], and other defect centers of small clusters have also been described [17].

Crystallites prepared by subliming NaCl *in vacuo* yield cubes typically 100 nm on a side, as transmission electron microscopy (TEM) examination shows [18]. At the resolution possible with the TEM images, this collection of salt particles again resembles that of Fig. 1, except that the cuboid crystallites are many orders of magnitude smaller.

Another important form of salt is that found in the atmosphere. Sea salt aerosols, generated by wave action over the earth's oceans, are among the most abundant particulate masses of the atmosphere [8]. The sea salt particles initially formed are droplets with diameters in the micron range, but can be converted to crystallites as winds drive them aloft to regions of lower relative humidity [19]. SEM of sea salt particles shows hollow many-faceted crystalline structures, rather resembling miniature geodes [20]. For the practiced eye, these particles look like aggregates of cubes, resembling those shown in Fig. 1.

In a number of studies on NaCl samples prepared in the laboratory [21–24], it has become apparent that adsorbed water on the surface of salt plays an important role in its heterogeneous chemistry. We should also anticipate, from the preceding discussion, that defect sites as well as smooth face sites of the salt surfaces will be important in determining the properties of adsorbed water.

Smooth NaCl (001) faces can be easily prepared in the laboratory by cleavage of a single crystal boule with a light tap on a chisel. Cleaved NaCl crystals revealing (001) faces are shown in Fig. 2. As can be seen in the photograph, the faces are not, however, entirely smooth, but typically consist of terraces interrupted by steps of less than a millimeter. These crystals look qualitatively similar to the table salt grains in Fig. 1, but are of course many orders of magnitude larger. Magnified images of a NaCl (001) face show it to be remarkably smooth in selected regions. Gold decoration techniques followed by

Fig. 2 Single crystals of NaCl. The crystals 1 to 2 cm on a side have been cleaved to expose (001) faces. Photograph by James Ewing

TEM examination [25–27] or atomic force microscopy (AFM) [28–31] reveal that the smooth terraces extend 0.1 μm to 1 μm or so before being interrupted by steps of atomic dimensions. An example of a TEM image of NaCl (001) is shown in Fig. 3 from the work of Hucher et al. [26]. That the laboratory surfaces generated by careful cleaving have a low density of defects is also made evident by the clear diffraction pattern observed in helium atom scattering (HAS) experiments [32]. However, the NaCl surface is negatively charged after cleaving [33, 34], suggesting an imbalance among cation and anion vacancy defect sites at the surface [35].

The discussion of water adlayer structures on NaCl surfaces can be naturally divided into two general areas that are tied to a temperature range. In the *low temperature* regime, we will be talking about theoretical calculations where the arrangement that water molecules assume on an NaCl surface is effectively at 0 K. The structures calculated are essentially timeless. For ultrahigh vacuum (UHV) experiments, cryogenic temperatures are required

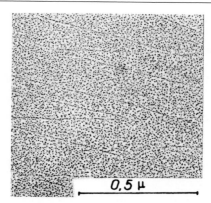

Fig. 3 Transmission electron microscopy of NaCl (001). From Hucher et al. [26]

to prepare water adlayers at pressures typically near 10^{-8} mbar [36]. At these low temperatures and pressures, the lifetime of an adsorbed water molecule is perhaps minutes or even hours. Therefore, low temperature water adlayers on NaCl are essentially static. By contrast under *ambient temperatures*, say 0–30 °C, the water vapor pressure is in the order of 10 mbar and the lifetime of a molecule within the adlayer is more likely in the microsecond scale [37]; its specific arrangement within the adlayer may be even orders of magnitude shorter [2]. So ambient conditions find the water adlayers or films essentially dynamic and ever changing into different structures.

It is for these reasons that the following sections distinguish low temperature and ambient temperature water adlayers on NaCl.

3
Low Temperature Water Adlayers

3.1
Single H$_2$O on NaCl (001)

The laboratory experiment to study a single H$_2$O molecule on NaCl (001) has yet to be performed, so we will rely on theoretical approaches. Here, we are talking about a lone water molecule at the global minimum of its intermolecular potential as defined by its position over an optimal site of the ionic substrate surface. Using a Lennard-Jones potential and distributed point charges, as well as semiempirical calculations or *ab initio* methods, a large number of theoretical investigations have been directed to explore the bonding of H$_2$O to NaCl (001) [36, 38–48]. Since, as Enkqvist and Stone show [41], the adlayer structures of H$_2$O on NaCl (001) are particularly sensitive to the potential chosen they have carefully developed its form using intermolecular perturbation theory (IMPT). For the water-NaCl surface, repulsion parameters were developed using IMPT, and dispersion coefficients were calculated using coupled Hartree-Fock perturbation theory. One advantage of this involved approach is that the contributions from various energy terms may be assessed. In their analysis of a single H$_2$O molecule on NaCl (001), they identify four major contributions to its bonding energy. Their calculations find the electrostatic energy, which includes dipole and higher order distributed multipoles, to be – 57 kJ mol^{-1}, repulsion energy at + 43 kJ mol^{-1}, induction energy of – 13 kJ mol^{-1}, and a dispersion energy of – 13 kJ mol^{-1} for a net binding energy of – 40 kJ mol^{-1}. It is not surprising that the electrostatic term dominates this assay because of the strong electric field arising from the ionic charges of the substrate. The Coulomb attraction of the dipole of a water molecule next to an isolated Na$^+$ ion is – 96 kJ mol^{-1} [2]. The compensating repulsion of Cl$^-$ ions in the substrate along with the appropriate Ewald summation over all the lattice charges, together with consideration of higher multipoles and most favorable molecule position, lowers the electrostatic attraction to – 57 kJ mol^{-1}. That the induction energy has a rather large contribution is again the consequence of the ionic substrate that can induce a large dipole and other multipoles in water drawing the molecule to the surface. Perhaps the most surprising result in the energetics calculation is the significant dispersion contribution. If we consider H$_2$O and Na$^+$ isoelectronic to Ne and Cl$^-$ isoelectronic to Ar, the dispersion contribution to the two-

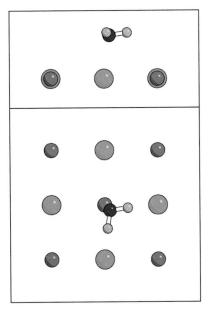

Fig. 4 Single H$_2$O on NaCl (001). Side view (*upper panel*) and top view (*lower panel*). The *smaller surface spheres* are Na$^+$ ions, the *larger surface spheres* are Cl$^-$ ions. From Engkvist and Stone [41]

body $H_2O \cdots Na^+$ or $H_2O \cdots Cl^-$ interactions should resemble that of a van der Waals complex Ne_2 or NeAr, giving an attractive energy of the order of $-1\,kJ\,mol^{-1}$ [49]. However, the dispersion attraction of a water molecule on NaCl (100) involves the many nearby ions at the surface and those of the substrate just below [2]. This accounts for the order of magnitude increase in the dispersion attraction of H_2O on a substrate over that in a van der Waals molecule. All insulator substrates, ionic or not, will contribute significant dispersion attractions to water adsorption. Repulsion terms will also be present in the energy considerations of water structures on all substrates. The general features of the single H_2O on NaCl(001) arrangement derived by Engkvist and Stone and shown in Fig. 4 are not surprising. The negative, or oxygen end of the water molecule, is poised near the Na^+ ion and its positive hydrogens, tilted slightly away from the NaCl(001) surface, are directed toward Cl^- ions. This structure is qualitatively similar to the ones proposed by others [32, 44].

3.2
Clusters and Monolayer H₂O on NaCl (001)

Ultrahigh-vacuum (UHV) conditions are required for the use of surface-sensitive techniques such as low-energy electron diffraction (LEED) and helium atom scattering (HAS) that can explore the architecture of the water adlayer on NaCl (001). The low water vapor pressures needed, less than about 10^{-8} mbar, in turn demand temperatures below $-160\,°C$ [36]. Consequently, only solid-like adlayers are explored. Fölsch et al. [50] have used LEED to study the water structure on the NaCl (001) surface. The scattering pattern they observed was consistent with an adlayer structure of a well ordered ice-like $c(2 \times 4)$ bilayer. This structure is similar to molecularly thin ordinary I_h ice, except that the adsorbed bilayer is slightly distorted to achieve compatibility with the NaCl (001) lattice. This adlayer then contains two types of water molecules. The first is in the bottom half of the layer, where the electronegative oxygen atoms in the water molecules are electrostatically attracted to Na^+ ions on the (001) surface and the hydrogen atoms attached to water molecules in the top half. Water molecules in the top half of the layer are not in direct contact with the substrate. Bruch et al. [36] studied adsorption of water to NaCl (001) by HAS under similar cryogenic temperatures and very low pressure conditions. In their analysis, the NaCl (001) lattice accommodates a (1×1) water monolayer with one H_2O for each Na^+Cl^- surface ion pair. To arrange the dipoles favorably, they suggested that the water molecules lie flat against the surface. While the LEED and HAS investigations indicated an ordered 2D condensed phase, IR (infrared) interrogation of D_2O on NaCl (100) at temperatures near 150 K and 10^{-9} mbar of water vapor showed no evidence for such a phase [51].

Theoretical methods have also been used to determine the adlayer structure. Wassermann et al. [38] used molecular dynamics calculations to explore the orientation of the water dipole moment as a function of coverage. According to their calculations, for low coverages, the water dipoles are perpendicular to the surface. Beyond half a monolayer they found that the dipoles begin to tilt toward the surface to favor hydrogen bonding between water molecules. Their theoretical adlayer structure agrees with the $c(2 \times 4)$ bilayer structure found by Fölsch et al. [50]. Calculations by Bruch et al. [36] and Jug and Geudtner [40] tend to favor a (1×1) layer with each molecular dipole (i.e. the H$_2$O molecular axis) nearly parallel to the surface.

In their theoretical approach to explore water clusters and monolayer structures on NaCl (001), Engkvist and Stone [41] have chosen the ASP-W4 water-water interaction potential. This has been shown to reproduce energies and structures of the gas phase water dimer as well as clusters accurately [52]. The ASP-W4 potential together with an accurate water-NaCl potential has allowed the exploration of a variety of water clusters on NaCl (001). While the water adsorbed as a dimer shows no indication of hydrogen bonding, the trimer and tetramer clusters do exhibit hydrogen bonding structures. A one dimensional water chain favors hydrogen bonding as evidenced by the molecular arrangements and the increase in bonding energy from the adsorbed

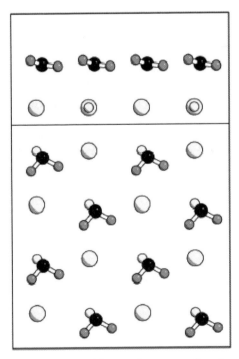

Fig. 5 Monolayer H$_2$O on NaCl (001). Side view (*upper panel*) and top view (*lower panel*). The *smaller white spheres* are Na$^+$ ions, the *larger white spheres* are Cl$^-$ ions. From Engkvist and Stone [41]

monomer at $-40\,\mathrm{kJ\,mol^{-1}}$ to $-54\,\mathrm{kJ\,mol^{-1}}$ (per water molecule). They also explored five monolayer structures. The energies range from $-48.4\,\mathrm{kJ\,mol^{-1}}$ (per water molecule) for the (1×1) structure shown in Fig. 5 and favored by Bruch et al. [36], to $-54.7\,\mathrm{kJ\,mol^{-1}}$ for the $c(4 \times 2)$ structure determined by Fölsch et al. [50]. A possible reason for the discrepancies between the two low temperature structures determined by UHV experiments may lie in the way in which the NaCl substrates were prepared. For the HAS studies, in-situ cleaved NaCl (001) single crystals were used [36], while epitaxially grown NaCl adlayers on a Ge(100) surface were employed in the LEED investigation [50]. With such a small difference in energy in the calculated (1×1) and $c(4 \times 2)$ adlayers, any subtle differences in the substrate might affect which structure is favored. Although the disagreement about the structure of water on NaCl (001) at low temperatures has not been resolved, we will see that these proposed structures need not have much bearing on the H_2O adlayer arrangement under ambient conditions to be considered in Sect. 4.

3.3
H$_2$O on Undefined NaCl Surfaces

The surfaces obtained by cleaving single crystals of NaCl are clearly (001), and a variety of techniques show them to be remarkably defect-free. However, the nanocrystallites produced by sublimation methods, while they appear cubic, their surfaces are poorly defined [18]. The diffuse infrared bands of CO adsorbed to these crystallite faces is suggestive of the heterogeneity of the surfaces [53]. Water adsorbed onto NaCl crystallites yields infrared band shapes [54, 55] quite distinct from those we shall discuss in a later section for thin film water on defined NaCl (001) faces.

One characteristic response of an assembly of NaCl crystallites on introduction of water vapor is that they sinter at low coverages [55]. A more spectacular response is that they exhibit chemical reactivity. Introduction of water vapor results in its reaction with surface defects to produce OH^- ions that become incorporated in the crystallites [55]. Mechanisms for this chemistry are discussed by Barnett and Landman [17].

4
Ambient Temperature Water Adlayers

4.1
Early Work

In the previous discussion of adlayers of water at low temperatures, the expectations and indeed the findings are that the molecules are rigidly bound to the surface. In principle and often in practice, we can know the structure or arrangement of the H_2O molecules. Under ambient conditions the adlayer may be liquid-like, sometimes described as a quasi-liquid layer [22, 24], suggesting that molecular orientations and positions can only be roughly specified. We then anticipate that ambient water layers or films will be amorphous.

The study of thin water films on insulators started, as did many of the pioneering investigations in surface science, with Irving Langmuir [1]. In 1918 he measured water coverage on mica and glass. His procedure, elegant in its simplicity, involved taking many sheets of mica, or cover glass slides, from the ambient laboratory environment and stacking them into a small vial. The adsorbed molecules (principally H_2O) on these surfaces were driven off by heating to 300 °C and captured in a trap cooled with liquid air. The number of water molecules caught, together with the known geometric area of the substrate surfaces, allowed a calculation of thin water film coverages: two molecular layers on mica and four on glass. If we view these insulator substrates as typical, then we come to expect any insulator surface to have some water molecules stuck to them under ambient conditions.

Work since Langmuir on water adlayers has followed two distinct paths. With the development of UHV technology, thousands of studies of water on metal and non-metal surfaces have been performed [56, 57]. Typically, water molecules on well-defined surfaces are locked into an ordered structure for long times (hours). The stability and order are dictated by the strength of the adlayer bond or the low temperature of the substrate as we have seen. The path less traveled for water adlayer studies is for exploration of ambient thin films. With equilibrium pressures in the mbar pressure range, the scores of surface interrogation techniques, HAS, LEED, X-ray photoelectron spectroscopy, etc., that depend on the low background pressures of the UHV chambers fail for water thin film studies. However, four particularly successful general approaches have recently been directed toward the investigation of thin film water on NaCl: transmisson electron microscopy (TEM), intermolecular force measurements, e.g. AFM, molecular simulations, e.g. molecular dynamics (MD) and Monte Carlo (MC), and infrared spectroscopy. We shall now review the findings of these approaches.

4.2
Thin Film Water on NaCl (001)

One of the earliest studies of the room temperature water adlayer on NaCl (001) was performed by Hucher et al. [26] using TEM. These images of the dry cleaved surfaces revealed smooth (001) terraces ~ 0.1 μm wide interrupted by steps of atomic dimensions as shown for example in Fig. 3. Hucher et al. then explored the morphological changes by exposing the crystals to water vapor, waiting for equilibrium to be established, evacuating the system, decorating with a thin gold deposit, and then interrogating the stabilized surfaces by TEM. After exposure to 20% RH, their microscopy revealed that the occasional sharp intersections of steps on the surface became rounded, suggesting that, at low pressure, water tends to favor adsorption to these corner sites and induce annealing. After a NaCl (001) face remained in an environment of 42% RH, the TEM image showed that the concentration of steps on the surface increased abruptly, suggesting that some sort of step movement had occurred.

Hucher et al. [26] also used an independent experimental approach to explore surfaces changes in response to water exposure. Electrical conductivity was measured between two strips of silver painted on an air cleaved NaCl (001) face. Three distinct regions of surface conductivity were found that describe the development of the water adlayer. Below 40% RH, the first region exhibited an exponential growth in conductivity with pressure. They proposed that H^+ ions in the adsorbed layer act as current carriers. Between 40% and 50% RH, a more gentle increase in conductivity was measured. In this second region, solvated Na^+ ions were suggested to contribute to the surface conductivity. Beyond 50% RH, the conductivity rose abruptly with both Na^+ and Cl^- ions proposed to act as charge carriers. In these elegant experiments performed nearly 40 years ago, the first direct evidence that water radically alters the surface of NaCl (100), even under somewhat dry conditions, was documented.

Beginning in 1990, Meyer and Amer applied AFM to interrogate the surface of NaCl (001) [28]. Schuger et al. [29] identified steps and other defects on the surface and suggested that adsorbed water on the AFM cantilever tip may alter the (001) face. Shindo et al. [30, 58, 59] quantified the migration of steps as a function of relative humidity. Unlike the TEM measurements, which effectively fix the system after exposure to water by pumping out the vapor and coating the surface with gold, AFM interrogation can follow step movements in real time. Shindo et al. [59] for example observed no step migration below 40% RH, but found movement gradually ($\sim 1 \, \mathrm{nm \, min^{-1}}$) setting in at 47% RH and precipitously ($> 10 \, \mathrm{nm \, min^{-1}}$) near 65% RH. These humidity markers for surface changes are qualitatively consistent with the observations of Hucher et al. [26].

Salmeron and his group, realizing the complication introduced by adlayer water on the AFM tip, have developed a technique they call scanning polarization force microscopy (SPFM) [60, 61]. In this form of noncontact AFM tip imaging, the metalized AFM tip is electrically biased and responds to the dielectric properties of the water film and any accompanying ions. The method requires the tip to be of the order of 10 nm from the film, so the higher resolution of contact AFM is lost. On the other hand, the electrical properties of the film can be explored. An example of their investigation of thin film water on NaCl (001) [34] is shown in Fig. 6.

The upper panel shows a SPFM topographical profile of a region of the NaCl surface at 30% RH. The arrow indicates a monoatomic step (0.26 nm) with a uniformly smooth terrace on the right then dropping through a multiple step of 6.5 nm to another smooth terrace. The smooth terrace on NaCl

(001) in this region is of the order of 1 μm wide. The lower panel of Fig. 6 shows the electrostatic contact potential of the same area for different humidity values. At 30% RH the electrostatic potential was + 75 mv over the terraces. The monoatomic step potential (+ 60 mv) and multistep potential (− 10 mv) were more negative than the terraces. On increasing the humidity the electrostatic potential smooths out and appears to approach a uniform value of about − 20 mv. In the terrace regions, the authors suggest that the initial negative values of the steps at low humidity is due to Cl⁻ solvation in the region of the steps. Increasing humidity causes the Cl⁻ solvation to spread out over the terraces. It is important to note that there is no step movement in these images. The single and multiple step positions do not change from 30 to 38% RH in Fig. 6. The authors report that the solvation is reversible for humidities less than 40%, but for high values step movement sets in and the surface is irreversibly changed.

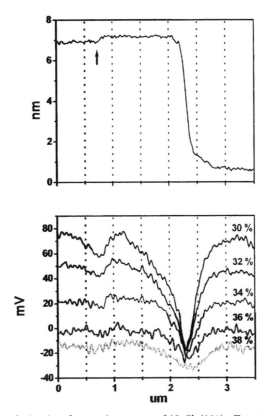

Fig. 6 Scanning polarization force microscopy of NaCl (001). *Top*: topographic profile showing a multiple step 6.5 nm high and a monoatomic step (marked by an *arrow*) at 30% RH. *Bottom*: contact potential of the same area for different humidity values. From Verdaguer et al. [34]

Interpretations of thin film water images using AFM or SPFM often suggest the presence of water and ions. However, the techniques cannot distinguish these species. With no adlayer water, i.e. an UHV experiment, Meyer and Amer identify Cl⁻ rather than Na⁺ only because the negative ions are so much bigger than the positive ions [28]. However, whether the migration of steps in the presence of water is accompanied by Na⁺ [62] or Cl⁻ [34], or H⁺ [26] is either speculation, or is supported by indirect measurements. While infrared spectroscopic interrogation cannot detect ions, it can unequivocally identify water.

Infrared interrogation of thin film water contains two important levels of information. The first is from the spectroscopic signature that can provide insight into the nature of the hydrogen bonding networks. Second, the extent of the spectroscopic response (absorption, reflection or extinction) yields an estimate of the film thicknesses for construction of isotherms and through them thermodynamic properties.

Before examining the infrared spectroscopy of thin film water, it is appropriate to review its ability to distinguish both phase differences and hydrogen bonding arrangements. In Fig. 7, taken from Ewing [63], we compare the optical cross sections, σ (cm^2 molec^{-1}), of liquid water at 27 °C [64] with that of ice at -7 °C [65] in the region of the ν_1 and ν_3 H$_2$O stretching vibrational modes. The large shift in the absorption maxima on the liquid to solid phase change, 3400 to 3200 cm^{-1}, is a consequence of different hydrogen bonding arrangements in these states. Normal mode analysis by Rice and coworkers [66] and more recently by Buch et al. [67], as well as qualitative considerations [68, 69], can account for some of these spectroscopic changes. The sharp feature near 3700 cm^{-1} is due to the dangling hydrogen (d-H) vibration. This vibration is associated with a hydrogen of a water molecule at the vapor interface that does not participate in hydrogen bonding. We have used the (non-hydrogen bonded) gas phase ν_1, ν_3 oscillator strengths [70], scaled by one half, and assumed a bandwidth of 20 cm^{-1} in keeping with previous surface measurements [55, 71–73] to generate the d-H band. We note also that all water features throughout the infrared (500–4000 cm^{-1}) undergo significant changes on phase transitions and their optical constants are well known [64, 65].

Our infrared approach to the model system, thin film water on NaCl (001), initiated in 1997 by Peters and Ewing [74, 75], has provided considerable insight into thin films on well defined surfaces. For example, the absorption spectra at 24 °C, taken from the study of Foster and Ewing [76], is shown in Fig. 8 for a variety of H$_2$O gas pressures. The substrate surfaces were prepared by cleaving slabs from a single crystal boule of NaCl and exposing (almost) defect-free (001) faces. A closely spaced stack of 14 slabs was placed in a temperature controlled optical cell that was set into the sample compartment of a Fourier transform infrared (FTIR) spectrometer and then evacuated. The small spaces between the crystal slabs and the presentation of 28 faces minimized the interfering absorption by the vapor and maximized absorption by adsorbed water.

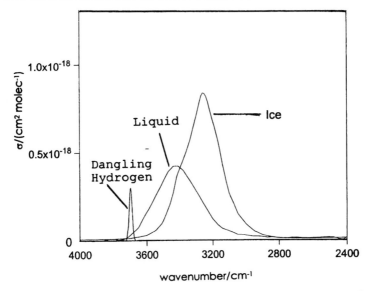

Fig. 7 Infrared optical cross sections for water. See text for details. From Ewing [63]

The series of spectra reveal a diffuse absorption associated with the vibrational stretching region of water molecules within the films on the NaCl(001) surfaces.

The close similarity of the liquid water profile in Fig. 7 with the 13 mbar spectrum of H$_2$O on NaCl(001) in Fig. 8 is good evidence that the thin film is liquid-like. Use of the optical constants of liquid water and the Beer-Lambert relation has enabled the determination of the coverage values as listed on Fig. 8 and the construction of the isotherm in Fig. 9. (A monolayer, $\Theta = 1$, corresponds to each Na$^+$Cl$^-$ surface ion pair covered, on average, by an H$_2$O molecule.) In order to provide a context with an ambient environment, the water pressure needed to produce a monolayer on NaCl(100) corresponds to 40% relative humidity [77, 78], a rather arid condition. In addition, the pristine appearing surfaces of the salt crystals in Fig. 2, photographed at 50% RH, were coated with several water layers.

The isotherm for water on NaCl(100) reproduced in Fig. 9 is rich in detail. Below a coverage of $\Theta = 0.5$, as suggested by Shindo et al. [59], two-dimensional islands form with H$_2$O molecules hydrogen bonded both to each other within each island and also to the surface. This bonding structure is consistent with the shift of the adsorption maxima to high wave numbers in the submonolayer spectra of Fig. 8. What appears to be a phase transition occurs between $\Theta = 1$ and $\Theta = 3$ where the hydrogen bonding network now extends between layers of molecules as the film thickens. The transition then is from a two-dimensional film to a three-dimensional film. Deliquescence, the spontaneous dissolution of NaCl at 23 mbar, is anticipated in the isotherm by the abrupt increase in coverage above $\Theta = 4$. (However, as we shall see

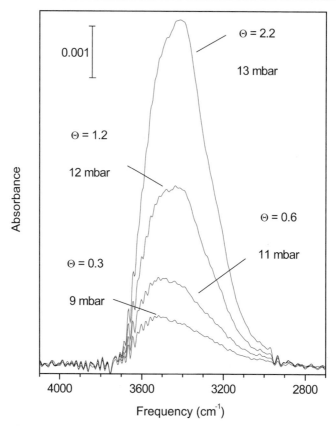

Fig. 8 Infrared spectra of thin film water on NaCl (001). The number of NaCl (001) crystal faces for these experiments, performed at 24 °C, was 28. From Foster and Ewing [76]

later, dissolution on NaCl (001) is a nucleated process and deliquescence occurs at a higher pressure.)

Foster and Ewing [76] have measured a family of isotherms. From the changing coverage values with pressure and temperature, they have been able to extract thermodynamic quantities (ΔF, ΔH, ΔS) that characterize thin film water for H_2O on NaCl(100). For example, the enthalpy of vapor condensation to form the monolayer film is $\Delta H = -50$ kJ mol^{-1} or more exothermic than for the condensation to liquid water that is -44 kJ mol^{-1} [79]. The monolayer film entropy is 15 JK^{-1} mol^{-1} lower than that of 70 JK^{-1} mol^{-1} for liquid water [79]. In summary, water molecules are more strongly bound to the NaCl(100) surface than in the liquid, and are more ordered.

The Monte Carlo calculations of Engkvist and Stone [46], based on the accurate intermolecular potential functions they have developed, allow a deeper understanding of the thin film molecular structure implied by the observed isotherms. They provide quantitative information in the form of pair distri-

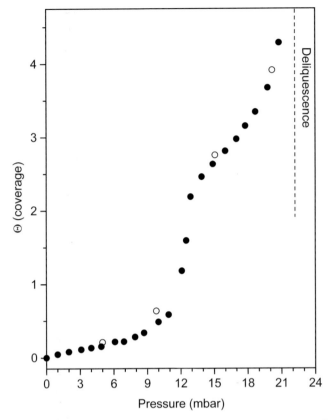

Fig. 9 An adsorption isotherm of water on NaCl (001). Data taken on ascending and descending pressures are given by *closed* and *open circles*, respectively. The temperature was 24 °C. Taken from Foster and Ewing [76]

bution functions, but first we shall draw from their calculations two configurations corresponding to coverages of $\Theta = 0.5$ and $\Theta = 3.0$. These are shown in Fig. 10. The exploded views into three distinct layers above the NaCl(001) surface, $0 \leq z \leq 360$ pm, $360 \leq z \leq 640$ pm, and $640 \leq z \leq 960$ pm, provide easier visualization of adlayer structures. These spacings have been guided by the layering revealed by the Monte Carlo distribution functions. Layer 1 for $\Theta = 0.5$ appears as a random assemblage of monomers and small clusters. However, the distribution functions actually indicate that the molecules are not randomly distributed but favor an $O - O$ separation consistent with hydrogen bond formation. Moreover, H$_2$O molecules are more likely over Na$^+$ than Cl$^-$ ions. Examinations of a number of Monte Carlo configurations suggest that molecules in layer 1 are rather anchored to the underlying ions. The few molecules in layer 2, some directly above molecules in layer 1, suggest the beginning of multilayer formation. Indeed two molecules are found in layer 3.

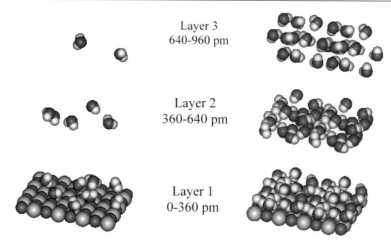

Layer 3
640-960 pm

Layer 2
360-640 pm

Layer 1
0-360 pm

Fig. 10 Exploded snapshots of H_2O on NaCl (001). The molecules are partitioned into three layers for coverages of $\Theta = 0.5$ (*left side*) and $\Theta = 3.0$ (*right side*). These images were taken from the calculations of Engkvist and Stone [46] and the figure from Foster and Ewing [76]

For a coverage of $\Theta = 3$, the water molecules are partitioned into distinct layers. This is clearly evident in the distribution function that explores molecular positions perpendicular to the (001) face. This layering of water, as revealed by the distribution of O centers above NaCl(001), is reminiscent of layering found for a liquid of hard spheres against a flat surface [2]. However the details of the distribution for the system – water molecules against the hydrophilic salt surface – are understandably more complicated. The molecules in layer 1 are now influenced by molecules in layer 2 as well as by the substrate below. Thus, while water molecules in the uppermost layer and the layer closest to the surface are only partially surrounded by other molecules, the distribution function that explores the O – O separation finds a value near that of liquid water. Thus, the adlayer $\Theta = 3$ resembles liquid water.

The pair distribution function calculated by Engkvist and Stone shown in Fig. 11 is a more quantitative representation than the "snapshots" shown in Fig. 10. The distribution for $\Theta = 0.5$ clearly shows a layer against the NaCl (001) face at an optimum separation near 0.26 nm (the sum of the Na^+ and H_2O radii [2]). However, the distribution also shows molecules spilling out beyond 1 nm. The distribution for $\Theta = 3.0$ shows three distinct layers with significant probability in regions expected for a fourth and higher layer. Engkvist and Stone also compare molecular distributions within the thin film calculated with their ASP-W4 potential with the simpler and more popular TIP4P potential. With TIP4P there is no evident layering

Shinto et al. [39] have performed a molecular dynamics simulation of water confined between pairs of (001) and (011) faces of NaCl using the

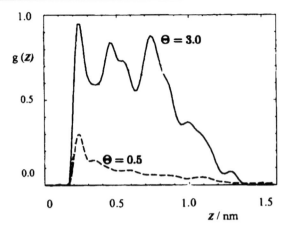

Fig. 11 Pair distribution functions for thin film water on NaCl (001). The distribution function, $g(z)$, is given as a function of the separations, z, of the water molecules from the NaCl (001) surface at coverages of $\Theta = 0.5$ and 3.0. Adapted from Engkvist and Stone [46]

SPC/E model. (For this potential, there are three point charges on each H₂O molecule and a Lennard-Jones potential between oxygen atoms on a pair of molecules.) Here only two layers are evident. Their calculations lead them to the conclusion that molecules near the (001) face are much more viscous that in bulk water. They use the term "solid-like" to describe the diffusive properties of water in the first two layers.

The theoretical modeling of both Engkvist and Stone, Shinto et al., and others have not included any Na⁺ or Cl⁻ ions that may have been solvated in the formation of thin film water. However, it is evident beginning with the work of Hucher et al. [26] and continuing with AFM studies [30, 31, 34, 58–62] that some ions are present in the thin films. Further theoretical work will need to include this fact.

4.3
Thin Film Water in and on Undefined NaCl Surfaces

Adsorption isotherms of water on powders (i.e. crystallites) of NaCl have been reported by Walter [80], Kaiho et al. [81], and Barraclough and Hall [82] at ambient temperatures. Since the methods for producing these powders (grinding, or precipitation from saturated salt solutions) must result in defect-rich surfaces, it is not surprising that the details of these isotherms differ from those in Fig. 9, although there are some qualitative similarities.

A series of papers originating in the laboratory of Finlayson-Pitts [22] and continuing in the group of Hemminger [24] describe water adlayers on NaCl as quasi-liquid. They, and others, demonstrate that this thin film plays a profound role in physical and chemical transformations of salt particles in the presence of traces of NO_2, HNO_3, SO_2 and other gases. This has particular relevance in atmospheric chemistry of naturally produced sea salt aerosols [22].

In an infrared study of laboratory produced NaCl aerosols, Weis and Ewing [83] offered models on the morphology of water in (or on) the salt particles. It has been known for a long time that salt crystals in mines can be easily deformed without fracturing [12]. Thin film water that somehow penetrates crystals of salt has been implicated in inducing its plasticity in a humid environment [84]. Finally, thin film water between grains of salt has recently been modeled to understand geological processes [85].

4.4
Deliquescence

We begin with the definition of *deliquesce*. The *Oxford English Dictionary* traces the verb to the Latin *de-* + *liquescere*: "to melt away, dissolve, disappear" [86]. We can imagine a grain of salt in a humid environment being transformed into a brine droplet. The thermodynamic condition for this phase transition is most simply obtained by measurement of temperature and vapor pressure of a saturated salt solution. At 25 °C this is 24 mbar [78]. With the vapor pressure of neat water at this temperature of 32 mbar [77], the deliquescence of NaCl corresponds to 75% RH.

An early and elegant measurement of the deliquescence of NaCl (and a number of other inorganic solids) was performed by Twomey [87] 50 years ago. He affixed a small particle of salt to a spider web in an environment where the relative humidity could be controlled. Under arid conditions, the photographic image of the NaCl sample resembled one of the grains in Fig. 1 with various surface faces intersecting at right angles to produce a number of sharp edges and corners. As the relative humidity was increased toward 75%, the salt particles start to become rounded. Beyond this relative humidity the salt solid, in taking up water, enlarged to become a drop clinging to the spider thread. There have been many measurements of the deliquescence of NaCl using a variety of techniques, as summarized in the recent review by Martin [88]. They are in general agreement in finding deliquescence at $75 \pm 2\%$ RH for 25 °C.

There is an important caveat in all the deliquescence studies of NaCl reviewed by Martin. The samples were polycrystalline, or if a single grain, their surfaces were irregular and not defined. Since it is known that surface curvature, steps and defects can effect interfacial properties [2, 3, 6, 26, 27, 44], Cantrell et al. [89] undertook to study deliquescence of cleaved NaCl (001) faces placed in a temperature and humidity controlled environment. The onset of dissolution (deliquescence) was monitored by the appearance of a thick film of brine as determined from its infrared spectrum. The infrared signatures of neat water and brine are quite distinct, so the film grown was brine and not water. Photometry was used to determine the brine thickness which was on the order of 1000 nm or orders of magnitude larger than the thin films that yield spectra like those in Fig. 8. The film thickness as a function of relative humidity is shown in Fig. 12 for a variety of samples. The onset of deliquescence, from 80 to 85% RH is significantly higher than the thermodynamic value of 75% RH. Cantrell et al. concluded from their work that the deliquescence on a largely defect-free NaCl (001) surface is nucleated.

Unlike most liquids, which can be substantially undercooled before they freeze, the majority of crystalline solids cannot be superheated [90]. This asymmetry in the first-order phase transition must be due to differences in the mechanisms responsible for the two processes. Once a liquid is undercooled, its associated crystal phase is thermodynamically favored, but the overall reduction in the free energy can only be accomplished by surmounting a large energy barrier. This barrier is the surface energy required to generate the interface between the embryonic fragment of the more stable solid phase and the surrounding liquid. The process of surmounting the barrier which divides the metastable, undercooled liquid from the stable crystal is known as nucleation [35, 91, 92]. Since melting except in rare cases is not a nucleated process, superheating of the solid is not possible [2].

Nothing in the preceding discussion is specific to freezing or melting. Efflorescence and deliquescence are the first-order phase transitions corresponding to freezing and melting for two component systems of a soluble salt and water. Efflorescence, the homogeneous crystallization from the solution always occurrs at an RH less than that over the saturated solution [88]; the

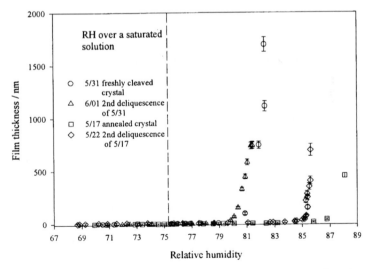

Fig. 12 Deliquescence of NaCl (001). Water film coverages are given as a function of relative humidity for NaCl crystals with different (001) surface preparations. From Cantrell et al. [89]

solution must be supersaturated for the phase transition to occur. Like freezing there is a large surface energy barrier to constructing an embryonic solid fragment in the surrounding liquid. Remarkably, the results of Cantrell et al. have demonstrated that both efflorescence *and* deliquescence are nucleated first-order phase transitions. The previous studies of deliquescence showing no apparent nucleation barriers are likely due to heterogeneous processes on the defect-rich polycrystalline samples. The nucleated phase transitions of salt solutions (efflorescence and deliquescence) are then somewhat analogous to the liquid-vapor phase transition where both condensation [92] and evaporation through boiling [93] are nucleated. The origin of the free energy barrier to deliquescence is actually well explored in the mechanism of pitting during water attack on soluble (or partially soluble) salts [94].

The density of defects affects the process of deliquescence because the energy required to solvate ions from the smooth (001) face differs from the energy required for ions associated with defects, as simple geometric considerations show. To pluck an ion from the (001) plane, the electrostatic attraction due to its five nearest neighbors must be overcome. In contrast, an ion at a step has one fewer nearest neighbors, and an ion at a corner has only three nearest neighbors. (Of course, the full summation of the energy of interaction between an ion and all the other ions in the crystal must be carried out to derive the full energetic barrier, but consideration of the nearest neighbors gives the correct trend.) The energy barrier associated with the deliquescence of ions from defects will be further reduced by the enhanced water-ion interaction that the defect affords. Quantum *ab initio* studies have shown that the adsorption energies of single water molecules on defects on the NaCl surface can be enhanced by 25% for F centers to almost 100% for steps [48].

The cleaved NaCl (001) faces have a low density of defects. The steps on NaCl (001) in Fig. 3 from Hucher et al. [26] are evident. Moreover, Verdaguer et al. [34] report that the NaCl (001) faces are electrically charged after cleaving suggesting point defects (vacancies or adions). It is not surprising that the mechanical shock of cleaving with chisel and hammer introduces a variety of surface defects. The onset of deliquescence variations found by Cantrell et al. from 80% RH to beyond 85% RH (see Fig. 12) suggests that defects play a role. Even the abrupt thick film growth beyond 85% RH may be defect induced, i.e. heterogeneous deliquescence. If one could prepare an annealed NaCl (001) face, freed of the trauma of the cleaving process, at what relative humidity would it deliquesce through a homogeneous process? And considering the results shown on Fig. 9, where film coverage below the equilibrium deliquescence relative humidity of 75% is greater than four molecular layers, certainly adequate for ion solvation, what role does thin film water play in the dissolution mechanism?

In concluding this section, it must be apparent that the mechanism of salt dissolution, on the molecular level, is poorly understood. It would seem to proceed at a number of levels. The first level must involve adatom or vacancy defects that will be present on any NaCl (001) face. These sites will be attractive to adsorbed water molecules either with chemical reaction [17, 95] or dissolution. The rounding of step intersections at 20% RH suggests dissolution of some sort [26]. The onset of step migration above 40% RH [26, 58, 59], but not below this value [34], indicates another level of dissolution. But none of these processes represents bulk dissolution which thermodynamics demands cannot occur until 75% RH. Whether the final stage of dissolution involves some sort of transformation of the step structure, pre-existing surface defects or the nucleated formation of pits [94] is not yet decided. In short, how salt dissolves in water is an open question.

Acknowledgements To begin, the contributions of Anthony Stone to this research are gratefully acknowledged. The collaborations between Stone and his group at Cambridge University and Ewing's group at Indiana University have been exceedingly enriching for over a decade. Discussions with Miquel Salmeron at Lawrence Berkeley National Laboratory have been particularly valuable. The contributions at Indiana University of Steve Peters, Michele Foster, Will Cantrell, Zhenfeng Zhang, David Dai, Rob Karlinsey, Charles McCrory, Vlad Sadtchenko, and Peter Conrad on various aspects of the study of thin water on salt surfaces are gratefully acknowledged. Thanks to you all. Finally the support of The National Science Foundation is acknowledged.

References

1. Langmuir I (1918) J Am Chem Soc 40:1361–1402
2. Israelachvili SN (1985) Intermolecular and Surface Forces. Academic Press, London
3. de Gennes P-G, Brochard-Wyart F, Quéré D (2002) Capillarity and Wetting Phenomena. Springer, New York
4. Faraday M (1991) Royal Institution Discourse, June 7, 1850; Experimental Researches in Chemistry and Physics. Taylor and Francis, New York
5. Elbaum M, Lipson SG, Dash JG (1993) J Cryst Growth 129:491
6. Adamson AW, Gast AP (1997) Physical Chemistry of Surfaces, 6th edn. Wiley, New York
7. Kirk-Othmer (1995) Encyclopedia of Chemical Technology, 4th edn. Wiley, New York
8. Warneck P (1988) Chemistry of the Natural Atmosphere. Academic, San Diego
9. Scharf D (1977) Magnifications, Photography with the Scanning Electron Microscope. Schocken Books, New York
10. (2003) Ullman's Encyclopedia of Industrial Chemistry, 6th edn. Wiley, Weinheim
11. Kittel C (1986) Introduction to Solid State Physics, 6th edn. Wiley, New York
12. Schmid E, Boas W (1968) Plasticity of Crystals. Chapman and Hall, London
13. Lennard-Jones JE, Dent BM (1928) Trans Faraday Soc 24:92
14. Benson GC, Freemand PI, Demsey E (1961) Adv Chem Ser 33, Am Chem Soc p 26
15. Bjorklund R, Spears K (1977) J Chem Phys 66:3437
16. Whetten RL (1993) Acc Chem Res 26:49
17. Barnett RN, Landman U (1996) J Phys Chem 100:13950
18. Zechima A, Sarano D, Garrone E (1985) Surf Sci 160:492
19. Blanchard DC, Woodcock AH (1980) Ann N Y Acad Sci 338:330
20. Cheng RJ, Blanchard DC, Chipaiano RJ (1988) Atmos Res 22:15

21. Hitchcock DR, Spitler LL, Wilson WE (1980) Atmos Envir 14:165
22. Vogt R, Finlayson-Pitts BJ (1994) J Phys Chem 98:3747; (1995) 99:13052
23. Peters SJ, Ewing GE (1996) J Phys Chem 100:14093
24. Allen HC, Laux JM, Vogt R, Finlayson-Pitts BJ, Hemminger JC (1996) J Phys Chem 100:6371
25. Bassett CA (1958) Philos Mag 3:1042
26. Hucher M, Oberlin A, Hobast R (1967) Bull Soc Fr Mineral Crist 90:320
27. Zangwill A (1988) Physics at Surfaces. Cambridge University Press, New York
28. Meyer G, Amer NM (1990) Appl Phys Lett 56:2100; (1990) 57:2089
29. Schuger AL, Wilson RM, Williams RT (1994) Phys Rev B 49:4915
30. Shindo H, Ohaski M, Baba K, Seo A (1996) Surf Sci 111:357–358
31. Dai Q, Hu J, Salmeron M (1997) J Phys Chem B 101:1994
32. Schmicker D, Toennies JP, Vollamn R, Weiss H (1991) J Chem Phys 95:9412
33. Harris LB, Fiasson S (1985) J Phys C: Solid State Physics 18:4845
34. Verdaguer A, Sasha GM, Ogletree DF, Salmeron M (2005) J Chem Phys
35. Frenkel J (1946) Kinetic Theory of Liquids. Dover Publications, New York
36. Bruch LW, Glebow A, Toennies JP, Weiss H (1995) J Chem Phys 103:5109
37. Atkins PA (1997) Physical Chemistry, 6th edn. WH Freeman, New York
38. Wassermann B, Mirbt S, Reif J, Zink JC, Matthias E (1993) J Chem Phys 98:10049
39. Shinto H, Sakakibara T, Higashitami K (1998) J Phys Chem B 102:1974
40. Jug K, Geudtner G (1997) Surf Sci 371:95
41. Engkvist O, Stone AJ (1999) J Chem Phys 110:12089
42. Park JM, Cho J-H, Kim KS (2004) Phys Rev B 69:233403
43. Meyer H, Entel P, Hafner J (2001) Surf Sci 488:177
44. Stockelmann E, Hentschke R (1999) J Chem Phys 110:12097
45. Garcia-Manyes S, Verdaguer A, Gorostiza P, Sanz F (2004) J Chem Phys 120:2962
46. Engkvist O, Stone AJ (2000) J Chem Phys 112:6827
47. Taylor DP, Hess WP, McCarthy MI (1997) J Phys Chem B 101:7455
48. Allouche A (1998) Surf Sci 406:279
49. Parson JM, Siska PE, Lee YT (1972) J Chem Phys 56:1511
50. Fölsch S, Stock A, Henzler M (1992) Surf Sci 65:264
51. Heidberg J, Häser W (1990) J Electron Spectrosc Relat Phenom 54/55:971
52. Hodges MP, Stone AJ, Xantheas SS (1997) J Phys Chem 101:9163
53. Ewing GE, Peters SJ (1997) Surf Rev Lett 4:757
54. Smart RStC, Sheppard N (1976) J Chem Soc Farad Trans II 72:707
55. Dai DJ, Peters SJ, Ewing GE (1995) J Phys Chem 99:10299
56. Thiel PA, Madey TE (1987) Surf Sci Rep 7/6–8:211–385
57. Henderson MA (2002) Surf Sci Rep 46/1–8:1–308
58. Shindo H, Ohashi M, Tateishi O, Seo A (1992) J Chem Soc, Faraday Trans 93:1169–1174
59. Shindo H, Okashi M, Tateishi O, Seo A (1997) J Chem Soc, Faraday Trans 93:1169
60. Hu J, Xiao X-D, Ogletree DF, Salmeron M (1995) Science 268:267; Hu J, Xiao X-D, Ogletree DF, Salmeron M (1995) Appl Phys Lett 67:476
61. Xu L, Lio A, Hu J, Ogletree DF, Salmeron M (1998) J Phys Chem B 102(3):540
62. Luna M, Rieutord F, Melman NA, Dai Q, Salmeron M (1998) J Phys Chem A 102:6793
63. Ewing GE (2004) J Phys Chem B 108:15953
64. Downing HD, Williams D (1975) J Geophys 80:1656–1661
65. Warren SG (1984) Appl Opt 23:1206–1225
66. McGraw R, Madden WG, Bergren MS, Rice SA, Sceats MG (1978) J Chem Phys 69:3483–3496
67. Buch V, Delzeit L, Blackledge C, Devlin JP (1996) J Phys Chem 100:3732–3744

68. Pimentel GC, McClelland AL (1960) The Hydrogen Bond. Reinhold, New York
69. Eisenberg DS, Kauzmann W (1969) The Structure and Properties of Liquid Water. Oxford University Press, New York
70. Ikawa SI, Malda S (1968) Spectrochim Acta 24A:655–665
71. Devlin JP, Buch V (1995) J Phys Chem 99:16534–16548
72. Horn AB, Chesters MA, McCustra MRS, Sodeau JR (1992) J Chem Soc, Faraday Trans 88:1077–1078
73. Becraft KA, Richmond GL (2001) Langmuir 17:7721
74. Peters SJ, Ewing GE (1997) J Phys Chem B 101:10880–10886
75. Peters SJ, Ewing GE (1997) Langmuir 13:6345–6348
76. Foster M, Ewing GE (2000) J Chem Phys 112:6817–6826
77. List RJ (ed) (1963) (Ed) Smithsonian Meteorological Tables, 6th edn. Smithsonian Institution, Washington D.C., p 347–352
78. Greenspan L (1977) J Res Natl Bur Stand, Sect A 81A:89
79. (1982) NBS Tables of Thermodynamic Properties. Published as J Phys and Chem Reference Data 11:Supplement 2
80. Walter HV (1971) Z Phys Chem 75:287
81. Kaiho M, Chikazawa M, Kanazawa T (1972) Nippon Kagaku Kaishi 1386
82. Barraclough PB, Hall PG (1974) Surf Sci 46:393
83. Weis DD, Ewing GE (1999) J Geophys Res 104:D17 21:275
84. Barnes RB (1933) Phys Rev 43:82; (1933) 44:898
85. Renard F, Orteleva P (1997) Geochim Cosmochim Acta 61:1963
86. The Oxford English Dictionary (1933, reprinted 1961) Clarendon Press, Oxford
87. Twomey S (1953) J Appl Phys 24:1099
88. Martin S (2000) Chem Rev 100:3403
89. Cantrell W, McCrory C, Ewing GE (2002) J Chem Phys 116:2116–2120
90. Dash J (1989) Contemp Phys 30:89
91. Temperley HNV (1956) Changes of State. Cleaver-Hume, London
92. Pruppacher HR, Klett JD (1997) Microphysics of Clouds and Precipitation. Kluwer Academic, The Netherlands
93. Zahn D (2004) Phys Rev Lett 93:227801
94. Teng HH (2004) Geochim Cosmochim Acta 68:253
95. Dai DJ, Peters SJ, Ewing GE (1995) J Phys Chem 99:10299

Struc Bond (2005) 116: 27–41
DOI 10.1007/430_003
© Springer-Verlag Berlin Heidelberg 2005
Published online: 1 November 2005

n-Body Decomposition Approach to the Calculation of Interaction Energies of Water Clusters

R. A. Christie (✉) · K. D. Jordan

Department of Chemistry, University of Pittsburgh, Pittsburgh, PA 15215, USA
haggis@pitt.edu, jordan@pitt.edu

Abstract A new methodology is proposed in which large basis set MP2-level calculations can be extended to water clusters with as many as 50 monomers. The computationally prohibitive scaling of traditional MP2 calculations is avoided by the use of an *n*-body decomposition (NBD) description of the cluster binding energy. The computational efficiency of the NBD approach is demonstrated by the application of the method in a Monte Carlo simulation of $(H_2O)_6$. Future development will further permit accurate MP2 calculations on clusters as large as $(H_2O)_{50}$.

1
Introduction

Water clusters have been extensively studied using both model potential and electronic structure methods. It is now well established that the MP2 electronic structure method, when employed with flexible basis sets, accounts in a near quantitative manner for the binding energies and relative stabilities of different isomers of water clusters. This conclusion has been established by carrying out large-basis set CCSD(T) calculations on the smaller clusters [1, 2].

The availability of the results of large basis set MP2 calculations on water clusters has provided considerable insight into the strengths and weaknesses of various model potentials for water [1–4]. Moreover, large or complete basis set (CBS) limit MP2-level binding energies for small water clusters are being exploited in the parameterization of a new generation of improved water models [5–8]. To illustrate the performance of three recently developed polarizable water models, we focus on the two low-energy isomers of $(H_2O)_{21}$ depicted in Fig. 1 (these results are from a recent study of Cui et al., unpublished data). Isomer **I** is the global minimum of $(H_2O)_{21}$ [9] as described by the TIP4P [10]

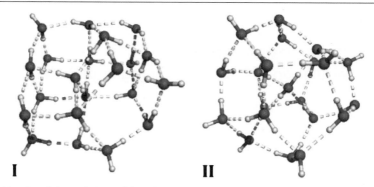

Fig. 1 Two low-lying minima of $(H_2O)_{21}$. Isomer **I** corresponds to the TIP4P global minimum as reported by Wales and Hodges [9], and isomer **II** consists of a dodecahedron-based $(H_2O)_{20}$ structure with an additional H_2O monomer inserted into the cage (unpublished data)

effective two-body potential, and isomer **II** can be viewed as being formed by inserting a water molecule inside a $(H_2O)_{20}$ water dodecahedron. In the latter case, two OH groups which protruded from the dodecahedral cage in the absence of the interior water molecule, have reoriented inwards, with the result that the encaged water molecule is engaged in a total of four hydrogen bonds.

Table 1 reports the binding energies of **I** and **II** calculated with the TIP4P model and with the Dang-Chang (DC) [11], TTM2-F [6], and AMOEBA [7] polarizable model potentials. In addition, results obtained at the B3LYP [12, 13]/aug-cc-pVDZ [14, 15], and RI-MP2 [16, 17]/aug-cc-pVTZ(-f) [14, 15] levels of theory are reported. Structure **II** is predicted to be 6.72 and 4.61 kcal mol^{-1} more stable than **I** at the B3LYP and RI-MP2 levels, respectively. Of the water models considered, only the AMOEBA model predicts **II** to be appreciably (3.98 kcal mol^{-1}) more stable than **I**. With the TTM2-F model, **II** is calculated to be only 1.14 kcal mol^{-1} more stable than **I**, and with the TIP4P and DC models, **I** is predicted to be about 1.0 kcal mol^{-1} more stable than **II**. It is clear from these results that correctly accounting for the relative

Table 1 Formation energies (in kcal mol^{-1}) for the two isomers of $(H_2O)_{21}$ shown in Fig. 1 (unpublished data)

Isomer	Theoretical method					
	TIP4P [a]	DC [b]	TTM2-F [c]	AMOEBA [d]	B3LYP [e]	RI-MP2 [f]
I	− 219.10	− 201.97	− 225.79	− 215.38	− 202.62	− 223.66
II	− 218.31	− 200.76	− 226.93	− 219.36	− 209.34	− 228.27

[a] [10] [b] [11] [c] [6] [d] [7] [e] B3LYP/aug-cc-pVDZ//B3LYP/aug-cc-pVDZ [f] RI-MP2/aug-cc-pVTZ(-f)//RI-MP2/aug-cc-pVDZ

energies of different isomers of a hydrogen-bonded cluster is an especially challenging task for model potentials.

The TTM2-F and AMOEBA models were parameterized using the results of MP2-level electronic structure calculations and are distinguished from most other water models by the use of three, atom-centred, mutually interacting polarizable sites. The TIP4P, DC, and TTM2-F models all employ three point charges to represent the static charge distribution of the water monomer, whereas the AMOEBA model employs atom-centred distributed multipoles up to quadrupoles. Thus, it appears that the key to the success of the AMOEBA model is its more realistic representation of the electrostatic potential of the water monomer.

Vibrational spectroscopy has proven especially important in the experimental characterization of water clusters [18–23]. Unfortunately, even the newer flexible (i.e. non-rigid monomer) water models do not fare well at accounting in a quantitative manner for the frequency shifts of the OH stretch vibrations of extended H-bonding networks. The problem here is well understood and stems from the limitation of generating flexible water models by simply grafting the force field of the monomer onto a rigid monomer model potential.

It is clear from the above discussion that, in spite of the steady improvements in water models, electronic structure calculations are still essential for addressing many problems of interest concerning isolated water clusters, as well as water clusters interacting with other species. Traditional MP2 calculations employing large basis sets are very computationally demanding. The largest MP2 calculations carried out to date on a water cluster are those by Fanourgakis et al., who reported the results of MP2/aug-cc-pVQZ calculations on $(H_2O)_{20}$ [4]. These calculations employed 3440 basis functions and were run using 512 processors on the PNNL Itanium 2 11.4 Teraflop supercomputer. On a more modest Beowulf cluster, single-point MP2/aug-cc-pVTZ calculations are feasible on clusters containing up to 20 or so water molecules. However, in those cases where it is necessary to locate a large number of stationary points, or to account for the properties of the clusters with excess internal energy or at finite temperatures, by performing Monte Carlo or molecular dynamics simulations, the large number of energy (and, possibly also, force) evaluations limits the calculations to rather small water clusters when using the traditional MP2 method with flexible basis sets.

These considerations lead naturally to interest in approaches for speeding up the electronic structure calculations. One way of accomplishing this is to adopt density functional theory (DFT), and, not surprisingly, density functional methods have been employed in many studies of water clusters (e.g. [24–27]). However, calculations with the common functionals, e.g. BLYP [13,28], and B3LYP [12,13], often fail to give accurate relative energies of different isomers, presumably due to the failure of commonly employed DFT methods to treat dispersion [29,30]. Recently, Xu and Goddard introduced the X3LYP density functional method [27]. Tests on small water clusters reveal that this functional performs better than BLYP and B3LYP at describing these clusters. However, for $(H_2O)_6$, the X3LYP functional, like the B3LYP functional, has an approximate 1.0 kcal mol^{-1} bias (as judged by comparison with complete basis set limit MP2 calculations) for the ring structure compared to the cage and prism isomers (unpublished data).

There are several options for reducing the computational costs of MP2 calculations. These include using localized-orbital MP2 (LMP2) [31–34], resolvent-of-the-identity MP2 (RI-MP2) [16, 17], and n-body decomposition (NBD) [3, 35] MP2 approaches. The LMP2 approach greatly reduces the CPU-time requirement and, when implemented using the Saebø-Pulay scheme [31–34] for choosing excitations, also reduces the basis set superstition error (BSSE) [36]. However, there is evidence that this approach introduces non-negligible errors in the interaction energies of water clusters [27]. The RI-MP2 method gives total energies very close to those from traditional MP2 calculations, while reducing the computational cost by a factor of 5–10. This approach was used in the above-mentioned investigation of isomers of $(H_2O)_{21}$ (unpublished data). However, it is still far too computationally demanding to be used in simulations of water clusters containing six or more monomers. The NBD procedure, which is discussed in detail below, has the potential of even greater computational speedups while retaining a highly accurate description of the energetics.

2
The n-Body Decomposition Procedure

The total binding energy of a cluster can be written as [3]

$$\Delta E = \sum_{n=1}^{N} E_n, \tag{1}$$

where E_n is the nth-body energy and N is the number of monomers in the cluster. The first three terms in the sum may be expressed as

$$E_1 = \sum_{i=1}^{N} \{E(i) - E_{\min}(i)\}, \tag{2}$$

$$E_2 = \sum_{i=1}^{N-1} \sum_{j=i+1}^{N} \{E(i,j) - E(i) - E(j)\}, \tag{3}$$

$$E_3 = \sum_{i=1}^{N-2} \sum_{j=i+1}^{N-1} \sum_{k=j+1}^{N} \{E(i,j,k) - E(i,j) - E(i,k) - E(j,k) + E(i) + E(j) + E(k)\}, \tag{4}$$

where $E(i)$, $E(i,j)$, and $E(i,j,k)$ are, respectively, the energy of monomer i, the (i,j) dimer, and the (i,j,k) trimer, with the geometries of the various fragments being extracted from that of the full cluster, and $E_{min}(i)$ is the energy of the monomer i at its relaxed structure. E_1 is thus the geometrical relaxation energy. Analogous expressions can be readily written down for the 4- and higher-body interaction energies.

The n-body decomposition procedure has been used in several studies to dissect the net interaction energies of water and other cluster systems into 1-body, 2-body, 3-body, etc., contributions [3, 35, 37, 38]. For water and other H-bonded systems, this expansion converges quite rapidly, with a typical accuracy ($0.05 \, \text{kcal mol}^{-1} \, \text{molecule}^{-1}$) being achieved by summing the contributions through fourth order [3]. Even when the series is truncated at the three-body contributions, the errors in the binding energies are typically less than $0.3 \, \text{kcal mol}^{-1} \, \text{molecule}^{-1}$ [3]. This result suggests that the n-body decomposition procedure can be used to design computationally efficient schemes for calculating MP2-level interaction energies of clusters.

The simplest approach to developing an NBD procedure for calculating energies of clusters is to truncate the expansion at either the 3- or 4-body terms. In the absence of symmetry, the use of the NBD procedure, including up to 4-body interactions, requires the evaluation of the energies of N monomers, $N(N-1)/2$ dimers, $N(N-1)(N-2)/6$ trimers, and $N(N-1)(N-2)(N-3)/24$ tetramers. (Hereafter the NBD procedure including interactions through n^{th} order will be denoted NBD-n.)

Table 2 reports the relative computational costs of evaluating the 2-, 3- and 4-body interaction energies for the aug-cc-pVDZ, aug-cc-pVTZ and aug-cc-pVQZ basis sets assuming that the electronic structure method used scales as N^4. Although computational costs are generally specified in terms of the number of basis functions, if the same basis set is employed for the different sized clusters, then the computational cost can be directly related to cluster size. Also, while traditional MP2 calculations formally scale as N^5, the RI-MP2 approach, as implemented in Turbomole [39] displays an approximately N^4 scaling with systems of this size. As seen from Table 2, MP2-level NBD-4 calculations would actually require more computational time than would the supermolecule calculation when characterizing clusters as small as $(H_2O)_{20}$ if the same basis set were used for evaluating all the interaction terms. If NBD-3 level energetics are adequate for the problem being addressed, then there would be a small computational savings over a supermolecule calculation. However, the above analysis is based on the assumption that the same level of theory, e.g. RI-MP2/aug-cc-pVTZ, is used for evaluating all the interaction energies. It will be shown below that it is possible to adopt more approximate procedures for evaluating the 3- and 4-body interaction energies than for the 2-body interaction energies, with little degradation of the net binding energies. Such an approach can lead to NBD-3 and even NBD-4 procedures that are significantly faster than supermolecule calculations for clusters the size of

Table 2 Comparison of the relative CPU time required for the calculation of various n-body components with the time necessary for a conventional supermolecule calculation of the energy of $(H_2O)_{20}$, assuming N^4 scaling

Basis set	n-body component	Relative time [a]
aug-cc-pVDZ	Supermolecule	1.6×10^5
	2-body	3.0×10^3
	3-body	9.2×10^4
	4-body	1.2×10^6
aug-cc-pVTZ	Supermolecule	4.1×10^6
	2-body	7.7×10^4
	3-body	2.3×10^6
	4-body	3.1×10^7
aug-cc-pVQZ	Supermolecule	5.0×10^7
	2-body	9.4×10^5
	3-body	2.9×10^7
	4-body	3.8×10^8

[a] One unit of time corresponds to that required to calculate the energy of a single H_2O molecule with the aug-cc-pVDZ basis set

$(H_2O)_{20}$. Moreover, as will be seen in the next section, in Monte Carlo simulations there is an additional factor of N advantage of the NBD-3 of NBD-4 procedures over supermolecule calculations.

Table 3 summarizes the n-body energies calculated at the MP2 level for three isomers of $(H_2O)_6$ (**III**, **IV**, **V** of Fig. 2), and Table 4 summarizes the n-body energies for four isomers of $H^+(H_2O)_5$ (**VI**, **VII**, **VIII**, **IX** of Fig. 3). Results are reported for both the aug-cc-pVDZ and aug-cc-pVTZ basis sets, with and without corrections for basis set superposition error (BSSE) [36]. For the ring form of $(H_2O)_6$, results are also reported for the aug-cc-pVQZ basis set. There are several conclusions to be drawn from the results summarized in Tables 3 and 4. We consider first the results for $(H_2O)_6$.

1. The 5- plus 6-body interaction energies are very small ($\leq 0.3\,\text{kcal mol}^{-1}$ in magnitude).
2. The 3- and 4-body interaction energies are sizable and are most important for the ring isomer, for which they are calculated to be about -11.6 and $-1.7\,\text{kcal mol}^{-1}$, respectively. This is a consequence of the reinforcing alignment of the dipoles of the monomers in the ring isomer.
3. With the aug-cc-pVDZ basis set, the BSSE corrections to the 3- and 4-body energies are as large as 0.9 and $0.4\,\text{kcal mol}^{-1}$, respectively. With the aug-cc-pVTZ basis set, the maximum BSSE corrections to the 3- and 4-body energies are reduced to 0.2 and $0.1\,\text{kcal mol}^{-1}$, respectively. Also, the BSSE-

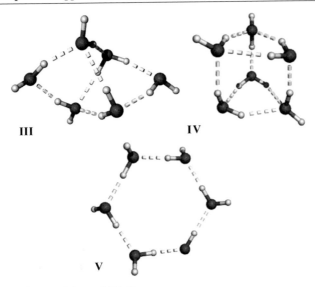

Fig. 2 Three low-lying minima of $(H_2O)_6$

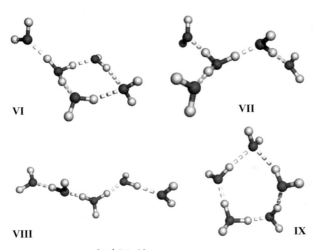

Fig. 3 Four low-lying minima of $H^+(H_2O)_5$

corrected 3- and 4-body energies calculated with the aug-cc-pVDZ basis set are nearly the same as those calculated with the aug-cc-pVTZ basis set.

4. The BSSE corrections to the 2-body energies are as large as 8.5, 4.3 and 2.1 kcal mol^{-1} with the aug-cc-pVDZ, aug-cc-pVTZ and aug-cc-pVQZ basis sets, respectively.

5. Although not included in Table 3, the $n \geq 3$ interaction energies calculated at the Hartree-Fock level are close to the corresponding MP2 results. This is not true of the 2-body interaction energies.

Table 3 MP2 n-body contributions to the energies (kcal mol^{-1}) of $(H_2O)_6$ minima **III**, **IV**, and **V** from Fig. 2. Reprinted from Pedulla JM, Kim K, Jordan HD (1998) Theoretical study of the n-body interaction energies of the ring, cage and prism forms of $(H_2O)_6$. Chem Phys Lett 291:78–84, with permission from Elsevier

n	Cage(III)			Prism(IV)			Ring(V)		
	DZ [a]	TZ [b]	QZ [c]	DZ	TZ	QZ	DZ	TZ	QZ
2-body	−41.05	−40.67	−40.21	−41.25	−40.86	−40.39	−35.03	−35.21	−34.91
	(−32.52) [d]	(−36.35)		(−32.83)	(−36.56)		(−27.75)	(−31.43)	(−32.84)
3-body	−8.17	−8.88		−8.23	−8.71		−11.24	−11.41	
	(−9.11)	(−9.13)		(−8.90)	(−8.90)		(−11.55)	(−11.60)	
4-body	−0.87	−0.60		−0.98	−0.66		−1.78	−1.82	
	(−0.48)	(−0.48) [e]		(−0.56)	(−0.56) [e]		(−1.73)	(−1.74)	
(5+6)-body	0.05	0.06		0.23	0.11		−0.21	−0.18	
	(0.01)	(0.01) [e]		(0.05)	(0.05) [e]		(−0.20)	(−0.19)	

[a] DZ denotes the aug-cc-pVDZ basis set
[b] TZ denotes the aug-cc-pVTZ basis set
[c] QZ denotes the aug-cc-pVQZ basis set
[d] Numbers in parenthesis correspond to counterpoise-corrected values
[e] These results have been estimated from the counterpoise-corrected 4+5+6-body interaction energies

Table 4 MP2 *n*-body contributions to the energies (kcal mol^{-1}) for isomers **VI**, **VII**, **VIII** and **IX** of H$^+$(H$_2$O)$_5$ from Fig. 3. Reproduced with permission from (2001) J Phys Chem A 105:7551–7558, Copyright 1998 Am Chem Soc

n	VI		VII		VIII		IX	
	aDZ [a]	aTZ [b]	aDZ	aTZ	aDZ	aTZ	aDZ	aTZ
1	2.39	2.89	2.88	3.36	6.90	7.60	6.73	7.45
2	−107.06	−107.56	−103.81	−104.49	−92.99	−94.19	−96.51	−97.39
	(−100.71) [c]	(−104.41)	(−97.89)	(−101.62)	(−86.81)	(−91.10)	(−89.59)	(−93.91)
3	11.59	11.69	9.39	9.36	−3.09	−3.00	0.10	0.12
	(11.69)	(11.71)	(9.47)	(9.44)	(−2.93)	(−2.94)	(0.10)	(0.15)
4	0.44	0.43	0.03	0.08	1.13	1.09	1.18	1.19
	(0.41)	(0.42)	(0.07)	(0.07)	(1.07)	(1.07)	(1.14)	(1.13)
5	−0.07	−0.05	−0.02	−0.03	0.02	0.03	0.06	0.03
	(−0.05)	(−0.05)	(−0.03)	(−0.03)	(0.02)	(0.03)	(0.05)	(0.05)
Net	−92.71	−92.60	−91.53	−91.72	−88.03	−88.47	−88.44	−88.60
	(−86.27)	(−89.44)	(−85.50)	(−88.78)	(−81.75)	(−85.34)	(−81.57)	(−85.13)

[a] aDZ denotes the aug-cc-pVDZ basis set
[b] aTZ denotes the aug-cc-pVTZ basis set
[c] Numbers in parenthesis correspond to counterpoise-corrected values

Most of the trends found in the *n*-body energies for $(H_2O)_6$ are also found for the isomers of $H^+(H_2O)_5$ shown in Fig. 3. Particularly noteworthy are the following observations:

1. The 5-body terms contribute at most 0.05 kcal mol^{-1} to the interaction energy at the MP2/aug-cc-pVTZ level.
2. The 4-body interaction energies are as large as 1.19 kcal mol^{-1}, with the largest value being for the ring isomer (**IX**). Nearly identical 4-body interaction energies are obtained with the aug-cc-pVDZ and aug-cc-pVTZ basis set, and BSSE makes negligible contributions to these 4-body energies.
3. The 3-body energy varies from 11.69 to –3.00 kcal mol^{-1}. This result contrasts with $(H_2O)_6$ for which the three-body energies are negative for the four isomers considered. The positive 3-body terms for some of the isomers of $H^+(H_2O)_5$ are a consequence of the impact of the proton on the arrangement of the water molecules in the network.
4. The 3-body energies obtained from the aug-cc-pVDZ and aug-cc-pVTZ basis sets are very close (agreeing to within 0.1 kcal mol^{-1}). Moreover, even with the smaller basis set, the BSSE corrections to the 3-body energies are very small.
5. The BSSE corrections to the 2-body terms are sizable even for a basis set as large as aug-cc-pVDZ.

It has been shown that near complete-basis set limit interaction energies of water clusters can be obtained by extrapolating MP2 energies from supermolecule calculations along the aug-cc-pVDZ, aug-cc-pVTZ, aug-cc-pVQZ and aug-cc-pV5Z sequence of basis sets [1, 2]. This approach is computationally prohibitive for water clusters much larger than $n = 10$. It is clear from the results reported in Tables 3 and 4 that the main reason that large basis sets are necessary to obtained converged binding energies using supermolecule calculations is the behaviour of the 2-body interactions. This result suggests a family of strategies for estimating MP2 level interaction energies with an NBD scheme. These include:

Scheme 1:
    1-body and 2-body – MP2/CBS
    3-body and 4-body – MP2/aug-cc-pVTZ
Scheme 2:
    1-body and 2-body – MP2/aug-cc-pVQZ
    3-body and 4-body – MP2/aug-cc-pVTZ
Scheme 3:
    1-body and 2-body – MP2/aug-cc-pVQZ
    3-body and 4-body – MP2/aug-cc-pVDZ with counterpoise correction
Scheme 4:
    1-body and 2-body – MP2/aug-cc-pVQZ
    3-body and 4-body – DC [11] or other polarizable model potential

Scheme 5:

      1-body and 2-body – MP2/aug-cc-pVQZ
      3-body – MP2/aug-cc-pVDZ with counterpoise correction
      4-body – DC or other polarizable model potential

In Scheme 1, CBS-limit 2-body energies would be obtained by extrapolation of the MP2 energies along the aug-cc-pVDZ, aug-cc-pVTZ, and aug-cc-pVQZ sequence of basis sets. In the other four schemes described above, the 2-body energies would be calculated at the MP2/aug-cc-pVQZ level, which would result in sizable errors in the net interaction energies. However, the errors in the relative energies of different isomers due to the use of MP2/aug-cc-pVQZ two-body energies are expected to be relatively small. Scheme 2, in which both the 3-body and 4-body interaction energies are calculated at the MP2/aug-cc-pVTZ level, would be almost as computationally demanding as the supermolecule calculations for a cluster the size of $(H_2O)_{20}$. However, if the 4-body interaction energies were neglected, the calculation would be about 20 times faster than the supermolecule calculation. Significant computational savings in the evaluation of the 3- and 4-body terms can be achieved by using the aug-cc-pVDZ basis set and applying the counterpoise correction for BSSE rather than using the aug-cc-pVTZ basis set (without counterpoise correction). With this strategy a NBD-4 calculation on $(H_2O)_{20}$ would be about 7 times faster than a supermolecule calculation, and a NBD-3 calculation would be about 360 times faster. Obviously, there are many other possible scenarios, depending on the magnitudes of the errors in the energies that can be tolerated. For example, 2-body interaction energies could be evaluated at the MP2/aug-cc-pVTZ level, with a more approximate method being used to evaluate the 3- and 4-body interaction energies.

Equally promising is the potential for reducing the computational effort required in evaluating 3- and 4-body interaction energies by screening based on the distances between the water monomers involved. Preliminary calculations on $(H_2O)_{21}$ (unpublished data), show that for this cluster, screening could lead to nearly an order of magnitude reduction in the CPU time required to calculate the 3- plus 4-body contribution to the interaction energy. In addition, it is possible to use smaller basis sets for calculating the long-range 2-body interactions than are used for the short-range 2-body interactions.

3
Monte Carlo Simulations of Water Clusters Using the NBD Procedure

Studies of the thermodynamic properties of small water clusters have been restricted thus far to the use of model potentials. Although it is possible to characterize the finite temperature behaviour of water clusters containing 20 or more monomers with density functional methods, the failure of current density functionals to treat dispersion interactions [29, 30] leads to inconsistencies when studying water cluster energetics (unpublished data). For this reason, it would be preferable to adopt instead wavefunction-based methods for the characterization of the thermodynamic properties of water clusters.

As noted in Sect. 1, the MP2 method has proven to be an accurate means of characterizing water clusters. However, the computational cost of traditional MP2 calculations has precluded the use of this method in Monte Carlo or molecular dynamics simulations. Even for a cluster as small as $(H_2O)_6$, at least 10^6 trial moves are required to obtain convergence of Monte Carlo simulations, and for larger clusters, the number of moves required to achieve convergence is generally much greater, even when using sampling schemes specifically designed to deal with quasi-ergodicity. These considerations have led us to develop an NBD-based algorithm for carrying out MP2-level Monte Carlo simulations of the finite temperature properties of water clusters [35]. The NBD scheme is ideally suited for carrying out such simulations since moves that involve only single molecule displacements require evaluation only of those interaction terms involving the displaced molecule. This reduces the number of 2-, 3- and 4-body terms that need to be evaluated for each move to $(N - 1)$, $(N - 1)(N - 2)/2$ and $(N - 1)(N - 2)(N - 3)/6$, respectively.

$(H_2O)_6$ has multiple isomers with different H-bonding topolgies that are close in energy, making this a good system to test the accuracy of the NBD method in simulations. In Sect. 2, it was noted that the interaction energy of $(H_2O)_6$ could be reproduced to an accuracy of 0.30 kcal mol^{-1} molecule^{-1} by employing an NBD scheme including up to 3-body interactions. In the study of Christie and Jordan [35], the NBD-3 procedure was used to carry out MP2-level Monte Carlo simulations of $(H_2O)_6$. The E_2 contributions in the simulations were evaluated at the LMP2 level of theory, whereas the Hartree-Fock method was used to calculate E_3. The simulations were carried out with the 6-31+ $G(d)$ [40–42] basis set as well as with the more flexible cc-pVDZ/aug-cc-pVDZ (for H/O) basis set, hereafter referred to as apVDZ.

The Monte Carlo simulations were carried out in the canonical ensemble, at temperatures of $T = 100$ and 220 K with the 6-31+ $G(d)$ basis set and $T = 220$ K with the apVDZ basis set. Each simulation was carried out for 1.1×10^5 trial moves, with the first 1.0×10^4 moves being used for equilibration.

In order to evaluate the accuracy of the NBD definition of the energy, supermolecule LMP2 calculations with the apVDZ basis set were carried out on 200 structures selected from the apVDZ Monte Carlo simulation trajectory. The average difference in the energies calculated with the NBD-3 and full-supermolecule LMP2 calculations, was found to be only 0.22 kcal mol^{-1}.

The average interaction energies $\langle E \rangle$, $\langle E_2 \rangle$, and $\langle E_3 \rangle$ from the apVDZ simulation are $- 31.29$, $- 26.98$ and $- 4.41$ kcal mol^{-1}, respectively. Thus the many-body component to the interaction energy is 14% of the total interaction energy of $(H_2O)_6$ at $T = 220$ K.

To shed further light into the nature of the $(H_2O)_6$ cluster at $T = 220$ K, structures saved from the apVDZ simulation were optimized to their local minima using the water model potential of Ren and Ponder [7]. The distribution of structures determined in this manner is similar to that found from simulations using the TIP4P [10] model potential at the same temperature [43]. Ring-based isomers (**V** from Fig. 2) are the most populated at the simulation temperature, accounting for 65.5% of the structures sampled. Book-based isomers account for 24.7% of the population, while the prism-based, cage-based and isomers of differing H-bond topologies account for less than 10% of the population.

The temperature dependence of several properties of $(H_2O)_6$ was calculated using the histogram method of Ferrenberg and Swendsen [44]. Using this approach, it was found that the total binding energy varies from $- 37.7$ kcal mol^{-1} at $T = 25$ K to $- 29.4$ kcal mol^{-1} at $T = 270$ K. The 3-body contribution ranges from 19.8% of the total interaction energy at $T = 25$ K to 12.9% at $T = 270$ K.

4
Conclusions

At present, the most commonly adopted method for obtaining accurate binding energies of water clusters is to carry out MP2 supermolecule calculations using a sequence of basis sets, such as aug-cc-pVDZ, aug-cc-pVTZ and aug-cc-pVQZ, and extrapolate to the complete basis set limit. This approach is very computationally demanding, being inapplicable to the study of clusters much larger than $(H_2O)_{20}$, and also could not be used to carry out MP2-level Monte Carlo simulations even on a cluster as small as $(H_2O)_6$. In this article we describe an *n*-body decomposition approach for addressing this problem. This approach exploits the finding that accurate interaction energies can be obtained by truncation of the *n*-body expansion at 4-body terms, and that one can use smaller basis sets for evaluating the 3- and 4-body energies than for the 2-body energies. For larger clusters, it should be possible to further reduce the computational effort by screening the individual 3- and 4-body components according to distance, and evaluating only those terms for which the distances between the monomers involved meet specified threshold values. In addition, further computational savings can be achieved by adopting more approximate procedures, e.g. using smaller basis sets for evaluating the long-range 2-body interactions than is required for the demanding short-range 2-body interactions. Effort is currently underway in our group to establish criteria for use in such screening procedures.

Interestingly, the *n*-body decomposition procedure also appears to be advantageous for dealing with the basis set superposition error problem. Specifically, calculations on small water clusters reveal that much of the error due to

BSSE arises from the 2-body interaction terms, and that this is much less of a problem for the 3- and 4-body interaction energies.

We have recently used the NBD-3 approach to carry out Monte Carlo simulations of $(H_2O)_6$. Although the basis sets used in that investigation were not large enough to obtain well converged energies, these simulations were carried out on a relatively small number of somewhat slow (1 GHz Pentium III) computers. By the use of a larger number of faster CPUs, NBD MP2 Monte Carlo simulations would be feasible on much larger water clusters. In particular, we note that the NBD procedure is well suited for Beowulf computer clusters because individual contributions to the various n-body terms can be run on different CPUs.

In summary, our initial application of the n-body decomposition procedure has proven quite promising. With further development, this approach should be able to provide near complete-basis limit binding energies for water clusters containing up to 50 monomers and enable Monte Carlo simulations to be run for water clusters containing up to 20 monomers.

References

1. Xantheas SS, Burnham CJ, Harrison RJ (2002) J Chem Phys 116:1493
2. Xantheas SS, Aprá E (2004) J Chem Phys 120:823
3. Pedulla JM, Kim K, Jordan KD (1998) Chem Phys Lett 291:78
4. Fanourgakis GS, Aprá E, Xantheas SS (2004) J Chem Phys 121:2655
5. Burnham CJ, Xantheas SS (2002) J Chem Phys 116:1500
6. Burnham CJ, Xantheas SS (2002) J Chem Phys 116:1515
7. Ren P, Ponder JW (2003) J Phys Chem B 107:5933
8. Kaminski GA, Friesner RA, Zhou R (2003) J Comp Chem 24:267
9. Wales DJ, Hodges MP (1998) Chem Phys Lett 286:65
10. Jorgensen WL, Chandrasekhar J, Madura J, Impey R, Klein ML J (1983) Chem Phys 79:926
11. Dang LX, Chang T (1997) J Chem Phys 106:8149
12. Becke AD (1993) J Chem Phys 98:5648
13. Lee C, Yang W, Parr RG (1988) Phys Rev B 37:785
14. Dunning Jr TH (1989) J Chem Phys 90:1007
15. Kendall RA, Dunning Jr TH, Harrison RJ (1992) J Chem Phys 96:6796
16. Weigend F, Häser M (1997) Theor Chem Acc 97:331
17. Weigend F, Häser M, Patzelt H, Ahlrichs R (1998) Chem Phys Letters 294:143
18. Pribble RN, Zwier TS (1994) Science 265:75
19. Huisken F, Kaloudis M, Kulcke A (1996) J Chem Phys 104:17
20. Buck U, Ettischer I, Melzer M, Buch V, Sadlej J (1998) Phys Rev Lett 80:2578
21. Brudermann J, Melzer M, Buck U, Kazimirski JK, Sadlej J, Bush V (1999) J Chem Phys 110:10649
22. Nauta K, Miller RE (2000) Science 287:293
23. Diken EG, Robertson WH, Johnson MA (2004) J Phys Chem A 108:64
24. Laasonen K, Parrinello M, Car R, Lee C, Vanderbilt D (1993) Chem Phys Lett 207:208
25. Lee C, Chen H, Fitzgerald G (1994) J Chem Phys 101:4472

26. Kim K, Jordan KD (1994) J Phys Chem 98:10089
27. Xu X, Goddard III WA (2004) J Phys Chem A 108:2305
28. Becke AD (1988) Phys Rev A 38:3098
29. Kristyán S, Pulay P (1994) Chem Phys Lett 229:175
30. Pérez-Jordá JM, Becke AD (1995) Chem Phys Lett 233:134
31. Pulay P (1983) Chem Phys Lett 100:151
32. Pulay P, Saebø S (1986) Theor Chim Acta 69:357
33. Saebø S, Pulay P (1987) J Chem Phys 86:914
34. Saebø S, Pulay P (1993) Annu Rev Phys Chem 44:213
35. Christie RA, Jordan KD (2005) Monte Carlo Simulations of the Finite Temperature Properties of $(H_2O)_6$. In: Dykstra C, Frenking G, Kim K, Scuseria G (eds) Theory and Applications of Computational Chemistry. Elsevier, New York, p 995–1009
36. Boys SF, Bernardi F (1970) Mol Phys 19:553
37. Xantheas SS (1994) J Chem Phys 100:7523
38. Christie RA, Jordan KD (2001) J Phys Chem A 105:7551
39. Ahlrichs R, Bär M, Häser M, Horn H, Kölmel C (1989) Chem Phys Lett 162:165
40. Hehre WJ, Ditchfield R, Pople JA (1972) J Chem Phys 56:2257
41. Clark T, Chandrasekhar J, Spitznagel GW, v R Schleyer P (1983) J Comp Chem 4:294
42. Frisch MJ, Pople JA, Binkley JS (1984) J Chem Phys 80:3265
43. Tharrington A, Jordan KD (2003) J Phys Chem A 107:7380
44. Ferrenberg AM, Swendsen RH (1988) Phys Rev Lett 61:2635

Struc Bond (2005) 116: 43–117
DOI 10.1007/430_004
© Springer-Verlag Berlin Heidelberg 2005
Published online: 18 October 2005

Intermolecular Interactions via Perturbation Theory: From Diatoms to Biomolecules

Krzysztof Szalewicz[1] (✉) · Konrad Patkowski[1,2] · Bogumil Jeziorski[2]

[1]Department of Physics and Astronomy, University of Delaware, Newark, DE 19716, USA
szalewic@udel.edu, patkowsk@physics.udel.edu

[2]Department of Chemistry, University of Warsaw, Pasteura 1, 02-093 Warsaw, Poland
jeziorsk@tiger.chem.uw.edu.pl

Abstract This article is devoted to the most recent, i.e. taking place within the last few years, theoretical developments in the field of intermolecular interactions. The most important advancement during this time period was the creation of a new version of symmetry-adapted perturbation theory (SAPT) which is based on the density-functional theory (DFT) description of monomers. This method, which will be described in Sect. 5.2, allows SAPT calculations to be performed for much larger molecules than before. In fact, many molecules of biological importance can now be investigated. Another important theoretical advancement was made in understanding the convergence properties of SAPT. It has been possible to investigate such properties on a realistic example of a Li atom interaction with an H atom. This is the simplest system for which the coupling of physical states to the unphysical, Pauli forbidden continuum causes the divergence of the conventional polarization expansion and of several variants of SAPT. This development will be described in some detail in Sects. 2–4, where, in addition to a review of published work, we shall present several original results on this subject. In an unrelated way, one of the most interesting recent applications of *ab initio* methods concerns the helium dimer and allows first-principle predictions for helium that are in many cases more accurate than experimental results. Therefore, theoretical input can be used to create new measurement standards. This broad range of systems that were the subject of theoretical investigations in recent years made us choose the title of the current review. With a few exceptions, the investigations of individual systems discussed here utilized SAPT. The calculations for helium are described in Sect. 6, recent wave-function based applications in Sect. 7, the performance of SAPT(DFT) on model systems in Sect. 8, and applications of SAPT(DFT) in Sect. 9. Section 10 summarizes work on biosystems.

1
Introduction

It is remarkable how broad a range of physical, chemical, and even biological phenomena originates from weak intermolecular interactions (also called van der Waals interactions), i.e. the interactions that do not involve forming a chemical bond between the interacting species. Intermolecular interactions (or forces) determine bulk properties of gases and liquids and are responsible for the very existence of molecular liquids and crystals. The knowledge of accurate intermolecular potential energy surfaces (PESs) is necessary to interpret high-resolution spectroscopic [1, 2] and scattering [3] data, including the spectroscopic data coming from planetary atmospheres and the interstellar gas [4], as well as to construct and tune empirical potentials used in

Monte Carlo (MC) or molecular dynamics (MD) bulk simulations [5]. In recent years, interactions of various molecules with helium became particularly important due to the development of superfluid helium nanodroplet spectroscopy [6, 7]. Weak intermolecular forces are responsible for the biomolecular recognition patterns and the catalytic activity of enzymes, and thus insights into intermolecular PESs are important for drug design [8, 9]. An interesting example of a macroscopic effect of van der Waals forces is given by the recent experimental evidence that Tokay geckos (*Gekko gecko*) owe their exceptional ability to climb smooth vertical surfaces to the van der Waals attractions between the surface and gecko toe-hairs [10].

For small monomers, the intermolecular potentials can be computed by standard electronic structure methods that account for electron correlation. This is done by using the *supermolecular approach*, i.e. for each configuration of fixed nuclei (the Born-Oppenheimer approximation), the interaction energy E_{int} is obtained by subtracting the total energies of monomers from the total energy of the cluster [11, 12]

$$E_{int} \equiv \mathcal{E} = E_{AB} - E_A - E_B . \tag{1}$$

The main advantages of the supermolecular approach are its universality, conceptual simplicity, and availability of many sophisticated *ab initio* methods and highly optimized computer codes that can be used to calculate the quantities on the r.h.s. of Eq. 1. However, this approach is not free from serious problems, mainly originating from the fact that the subtraction in Eq. 1 involves components that are several orders of magnitude larger than the interaction energy $\mathcal{E}$; in fact, the errors with which these components are calculated exceed the value of $\mathcal{E}$ for nearly all computations performed so far. Under these circumstances, the supermolecular approach may give accurate results only if a cancellation of errors occurs in Eq. 1. A necessary condition for this cancellation to take place is size-consistency of the method employed to calculate E_{AB}, E_A, and E_B [13]. However, this condition is not sufficient, as has been demonstrated by the failure of the supermolecular density functional theory (DFT) applied to several rare-gas dimers [14]. We now know that the cancellation of errors in Eq. 1 does take place if one uses Møller-Plesset perturbation theory (MP) or the coupled-cluster (CC) method to calculate the quantities on the r.h.s.; it is worth mentioning that this fact has been proved by a decomposition of the supermolecular interaction energy in terms of SAPT corrections [15]. However, even employing a highly correlated method like the supermolecular coupled-cluster with single, double and noniterative triple excitations (CCSD(T)) and large basis sets does not guarantee that an accurate PES will be obtained [16]. Another disadvantage of the supermolecular approach is that one must take care of eliminating the basis set superposition error (BSSE) [17, 18]. This requires extra computation time and, most importantly, it is not entirely clear how to get rid of BSSE when the partitioning of the dimer into interacting monomers is ambiguous, as in the case of PESs associated with chemical reactions, or when there exist multiple PESs resulting from the presence of an open-shell monomer [12]. Last but not least, all one can get from a supermolecular calculation at a given dimer geometry is a single number which tells nothing about the physics underlying the interaction phenomenon.

Another difficulty arising in computational investigations of intermolecular interactions is that in virtually all cases one has to include effects of electron correlation. The computer resource requirements of all methods involving electron correlation scale as a high power (5th or higher) of system size (as measured by the number of electrons), which severely limits the size of molecules that can be handled. A much faster approach is provided by the DFT method, which scales as the third power of the system size. DFT is widely used in solid state physics and in chemistry. Unfortunately, with the currently available functionals, DFT fails to describe an important part of intermolecular forces, the dispersion interaction. Consequently, predictions are poor except for very strong intermolecular interactions, as in the case of hydrogen-bonded clusters.

An alternative to the supermolecular approach is symmetry-adapted perturbation theory [19–21]. In SAPT, the interaction energy is computed directly rather than by subtraction. SAPT provides both the conceptual framework and the computational techniques for describing intermolecular interactions, including the dispersion energy. However, the computer resources required by SAPT, similar to those of the methods with high-level treatment of correlation used in the supermolecular approach, make applications to monomers with more than about ten atoms impractical at the present time. For ten-atom or smaller molecules, SAPT has been very successful; see [20, 21], and Sect. 7 for a review of applications.

The concept of calculating the interaction energy of two chemical systems A and B perturbatively is not at all a new idea. The first intermolecular perturbation expansion was proposed [22] just a few years after the foundations of quantum mechanics had been laid. Since then, numerous other expansions, now known under a common name of symmetry-adapted perturbation theory, have been introduced and the perturbation theory of intermolecular forces is now a fully mature approach. Thanks to the development of the many-body SAPT [23] and of a general-utility closed-shell SAPT computer code [24], the perturbative approach to intermolecular interactions has been successfully applied to construct PESs for numerous interacting dimers of theoretical and experimental interest [19–21, 25–27]. One of the notable achievements of SAPT is an accurate description of the interactions between water molecules [21, 28–32]. A recent paper by Keutsch et al. [33] compares the complete spectra of the water dimer with theoretical predictions obtained using an empirical potential fitted to extensive spectroscopic data, and with the predictions from a SAPT potential. These comparisons show that the latter potential probably provides the best current characterization of the water dimer force field. In another recent application, an SAPT PES for helium interacting with water has been used to calculate scattering parameters that agreed well with the high-quality experimental data [34, 35].

The SAPT interaction energy is expressed as a sum of well-defined and physically meaningful contributions, corresponding to the electrostatic, induction, dispersion, and exchange components of the interaction phenomenon. Thus, the SAPT theory provides the basic conceptual framework for our understanding of the nature of the intermolecular forces [19, 36]. Finally, one may note that SAPT is much more flexible computationally than the standard supermolecular approach. Different interaction energy components can be computed using different levels of the electron correlation treatment and/or employing different basis sets. One can also use custom designed basis sets, like the so-called monomer-centered "plus" basis sets (MC^+BS), to speed up the basis set convergence of some or all perturbation corrections [37].

2
Convergence Properties of Conventional SAPT

The question of the convergence of the SAPT expansions has been studied extensively in the past, but numerical investigations were possible only for very simple few-electron dimers like H_2^+ [38–42], H_2 [43–47], HeH [48], He_2 and HeH_2 [49]. Such systems do not exhibit all the complications arising in interactions of general many-electron monomers. For example, on the basis of early studies, it was believed for some time that the Hirschfelder-Silbey (HS) perturbation expansion [50, 51] should converge quickly for many-electron systems, just like it did for all the small systems investigated. Since low-order energy corrections of the extremely complicated HS method were almost identical to the ones calculated from the much simpler symmetrized Rayleigh-Schrödinger (SRS) theory [40], these results justified the use of the low-order SRS method in practical applications. However, neither of the systems mentioned above can be viewed as a legitimate model for studying the convergence of SAPT for many-electron systems, since significant complications appear when one of the interacting monomers has more than two electrons. These complications, first pointed out by Adams [52–55], arise from the fact that in such a case the physical ground state of the interacting dimer is buried in a continuous spectrum of unphysical states (discovered by Morgan and Simon [56]), which violate the Pauli exclusion principle. This situation is graphically displayed in Fig. 1 of [57] for a lithium atom interacting with a hydrogen atom. In the presence of the Pauli-forbidden continuum, the Rayleigh-Schrödinger (RS) perturbation theory, and thus also the SRS method which employs the same expansion for the wave function, must diverge. Moreover, under these circumstances the Hirschfelder-Silbey theory cannot be expected to converge either, since in this theory one performs a perturbation expansion of the so-called primitive [58] or localized [48] function, defined as a sum of all asymptotically degenerate eigenfunctions of the Hamiltonian, both physical and Pauli-forbidden. When the Pauli-forbidden eigenfunctions belong to the continuum, the primitive function is not square integrable and cannot be a limit of a convergent series of square integrable functions (all finite-order perturbed wave functions in the HS theory are well defined and square integrable).

There exists another class of SAPT expansions that is free from the Pauli-forbidden continuum problem and can be expected to converge for many-electron systems. This class includes the ELHAV theory – the one introduced in 1930 by Eisenschitz and London [22], and rediscovered later by Hirschfelder [59], van der Avoird [60], and Peierls [61]. Other SAPT expansions of this kind are the Amos-Musher (AM) [62] and Polymeropoulos-Adams [63] theories. These methods, however, suffer from a different problem, discovered in an early numerical investigation [43] of the H_2 molecule and confirmed in analytical studies [38, 64] of the H_2^+ ion: such methods fail to recover in the second order the important induction and dispersion components of the interaction energy, leading to wrong values of the constants C_n in the C_n/R^n asymptotic expansion of the interaction energy [65] (R denotes the intermonomer distance), starting from the C_6/R^6 term for electrically neutral monomers and C_4/R^4 term when at least one monomer is electrically charged. This failure should be contrasted with the behavior of the SRS and HS methods which are asymptotically compatible with the RS theory, i.e. have the property that each term in the $1/R$ expansion of the interaction energy is recovered in finite order [66]. On the basis of these findings, Jeziorski and Kołos proposed [67] a new SAPT expansion, the JK theory, which, while still being free from the Pauli-forbidden continuum problem, has the correct large-R asymptotics of the second-order energy (but not of the third- and higher-order corrections), thus reproducing correctly the conventional (second-order) induction and dispersion components of the interaction energy. However, none of the SAPT expansions discussed so far is simultaneously convergent for many-electron systems and asymptotically correct in every order of the perturbation theory. This fact has been elaborated by Adams [68], who concluded that the existing SAPT formulations are inadequate for the study of many-electron systems, and one must search for a new theory.

The understanding of the convergence issues described above has been greatly improved in the past few years. In particular, studies of the high-order convergence properties of the existing SAPT expansions for a system exhibiting the Pauli-forbidden continuum have been performed [57, 69]. Although the prediction of the divergence of several SAPT theories for many-electron systems was confirmed, it has been shown that these theories provide useful and accurate information despite their divergence. Other SAPT expansions have been found to converge in the presence of a Pauli-forbidden continuum, and their low-order convergence properties were greatly influenced by the asymptotic correctness/incorrectness of low-order energies. In particular, a qualitative relation was found between modifications in the symmetry-forcing procedure such as in the JK theory and improved asymptotics and convergence properties of SAPT [69]. Very recently, several new SAPT expansions were developed [70–72] that are simultaneously convergent in the presence of the Pauli-forbidden continuum and asymptotically correct to any order. These expansions are also accurate in low order for a wide range of intermolecular distances, and are therefore suitable for employing in practical calculations for many-electron systems. This development has been made possible by separating the singular, short-range part of the nuclear attraction terms in the interaction operator and treating it differently from the regular long-range part [70, 71]. The convergence properties of the resulting so-called "regularized expansions" will be discussed in detail in Sect. 3. In this section we shall review results obtained for the conventional SAPT formulations, treating the interaction operator as a whole.

All the methods discussed above have the property of being formulated in a completely basis set independent way. Another possible approach is to consider the Schrödinger equation in the matrix form using some specific basis set. If this matrix is decomposed appropriately, one can obtain a family of so-called "symmetric perturbation" treatments [19, 67]. The best known of these is the variant called intermolecular perturbation theory (IMPT) developed by Hayes and Stone, which has led to many successful applications in the many-electron context [73–76].

2.1
Polarization Approximation

Suppose one wants to calculate the interaction energy of two systems A and B which are, in the absence of interaction, described by the clamped-nuclei Hamiltonians H_A and H_B, respectively. Let ϕ_A and ϕ_B denote the ground-state eigenfunctions of H_A and H_B, respectively, and let E_A and E_B be the corresponding eigenvalues,

$$H_A\phi_A = E_A\phi_A , \quad H_B\phi_B = E_B\phi_B . \tag{2}$$

When the interaction between A and B is switched on, the dimer AB is described by the Hamiltonian $H = H_A + H_B + V$, where the operator V collects all Coulombic interactions between particles (electrons and nuclei) belonging to A and those belonging to B. For an eigenfunction ψ of the Hamiltonian H, the corresponding eigenvalue is equal to $E_A + E_B + \mathcal{E}$. Our aim is to calculate approximations to the interaction energy $\mathcal{E}$ and the dimer wave function ψ by means of the perturbation theory. The simplest way of doing so is provided by the conventional Rayleigh-Schrödinger (RS) perturbation theory with the zero-order Hamiltonian $H_0 = H_A + H_B$ and the perturbation equal to V [77] (in this context, the RS method is often called, after Hirschfelder [78], the *polarization approximation*).

To derive the equations for the RS perturbation corrections, it is convenient to rewrite the Schrödinger equation $H\psi = E\psi$ in the so-called Bloch form [79],

$$\psi = \phi_0 + R_0(\langle \phi_0 | V\psi \rangle - V)\psi \tag{3}$$

where $\phi_0 = \phi_A \phi_B$ is a (normalized) eigenfunction of the unperturbed Hamiltonian H_0, corresponding to the energy $E_0 = E_A + E_B$, and R_0 is the reduced resolvent of H_0 which can be defined by the formula

$$R_0 = (1 - P_0)(H_0 - E_0 + P_0)^{-1}, \tag{4}$$

with $P_0 = |\phi_0\rangle\langle\phi_0|$ being the projection onto the unperturbed function ϕ_0. One can easily verify that any ψ satisfying Eq. 3 fulfills the so-called *intermediate normalization* condition $\langle \phi_0 | \psi \rangle = 1$.

Equation 3 can be solved iteratively [36, 40, 67],

$$\psi_n = \phi_0 + R_0(\mathcal{E}_n - V)\psi_{n-1}, \tag{5}$$

where

$$\mathcal{E}_n = \langle \phi_0 | V\psi_{n-1} \rangle \tag{6}$$

can be viewed as the nth approximation to the interaction energy $\mathcal{E}$. When the iterative process is initiated with $\psi_0 = \phi_0$, the approximate energy $\mathcal{E}_n$ contains all the RS energy corrections $E_{RS}^{(k)}$ up to and including the nth order, as well as some terms of the order higher than n [67]. To extract the individual corrections $E_{RS}^{(n)}$, one needs to substitute $H \to H_0 + \zeta V$, where ζ is a complex variable, and insert the expansions

$$\psi(\zeta) = \sum_{n=0}^{\infty} \psi_{RS}^{(n)} \zeta^n, \quad \mathcal{E}(\zeta) = \sum_{n=1}^{\infty} E_{RS}^{(n)} \zeta^n \tag{7}$$

into Eqs. 5 and 6. The resulting expressions for $\psi_{RS}^{(n)}$ and $E_{RS}^{(n)}$ are

$$E_{RS}^{(n)} = \langle \phi_0 | V\psi_{RS}^{(n-1)} \rangle, \tag{8}$$

$$\psi_{RS}^{(n)} = -R_0 V\psi_{RS}^{(n-1)} + \sum_{k=1}^{n-1} E_{RS}^{(k)} R_0 \psi_{RS}^{(n-k)}, \tag{9}$$

where $\psi_{RS}^{(0)} \equiv \phi_0$.

Unlike the full Hamiltonian H, its zero-order part H_0, as well as the perturbation V, does not possess full symmetry with respect to the permutations of electrons. As a consequence, the zero-order wave function ϕ_0 is not completely antisymmetric – some electrons are assigned to one monomer, the remaining ones to the other. Therefore, the equations for the perturbed wave functions $\psi_{RS}^{(n)}$ (or the ψ_n functions of Eq. 5) cannot be solved in the Hilbert space $\mathcal{H}_{AB}$ of the Pauli-allowed (antisymmetric with respect to interchanges of any two electrons) functions of the dimer, but rather in a larger, product space $\mathcal{H}_A \otimes \mathcal{H}_B$, where $\mathcal{H}_A$ and $\mathcal{H}_B$ are the Hilbert spaces of Pauli-allowed functions for monomers A and B, respectively. At first glance, one could think that V amounts to a small perturbation of H_0, at least when the intermonomer distance R is large. However, this is not the case: the difference $||\phi_0 - \psi||$, where $||\cdot||$ is the L^2 norm in the Hilbert space, is always large and does not tend to zero when R goes to infinity [67]. The finding that V cannot be treated as a small perturbation has a dramatic manifestation: the spectral properties of the operators H_0 and H are completely different when one (or both) of the interacting monomers has more than two electrons. Furthermore, the lowest physical eigenstate of H lies in such cases within a continuum of Pauli-forbidden states (for a more detailed discussion of this issue we refer the reader to [52, 53, 57], and [72]). Under these circumstances, one must expect that the RS perturbation theory will have serious problems converging to the interaction energy $\mathcal{E}$.

Even if each of the monomers has two electrons or fewer, the polarization approximation, although convergent, is far from being suitable for practical applications. Large-order numerical studies for H_2^+ [39, 42] and H_2 [46] revealed that in low orders the sum of the polarization approximation approaches the so-called *Coulomb energy* Q, defined as an arithmetic mean of the energies of the lowest gerade and ungerade states (for H_2^+) or the lowest singlet and triplet states (for H_2). After the value of Q is reproduced to a good accuracy, convergence of the polarization series deteriorates dramatically and the remaining part of the interaction energy – the *exchange energy* – is not reproduced to any reasonable extent in finite order. The pathologically slow high-order convergence of the polarization expansion manifests itself in the values of the convergence radii ρ of the perturbation series being only marginally greater than unity (for instance, $\rho = 1.0000000031$ for H_2 at the van der Waals minimum distance of 8 bohr) [46]. For the helium dimer, the situation is even worse. Not only the sum of the polarization expansion converges extremely slowly after reaching the value of Q (see [80] for the definition of Q for He_2), but its limit is not the physical ground-state energy, as it was for H_2^+ and H_2, but rather the energy of the unphysical, fully symmetric $1(\sigma_g)^4$ state [49, 81].

When one of the monomers has three or more electrons, the polarization approximation diverges, as first proved by Kutzelnigg [58]. He argued that an avoided crossing must take place when the ground-state interaction energy $\mathcal{E}(\zeta)$ is analytically continued from the physical value at $\zeta = 0$ to the low-lying unphysical value at $\zeta = 1$. The existence of the unphysical continuum only makes matters worse: $\mathcal{E}(\zeta)$ has to be continued analytically through infinitely many avoided crossings before reaching the physically significant value of $\zeta = 1$ [55]. It is worth noting that it is the one-electron, attractive part of the perturbation V that is responsible for the existence of the avoided crossings and the divergence of the RS theory for many-electron systems. The computationally much more complicated two-electron part of V plays only a minor role in determining the convergence properties. An interesting model for studying the convergence of intermolecular perturbation series, based on the above observation, has recently been proposed by Adams [82]. This model neglects the electron-electron interaction completely; however, it can provide qualitative predictions of the convergence/divergence of the polarization approximation for various dimers. Adams' model fails in the case of a ground-state helium atom interacting with a ground-state hydrogen atom, for which it predicts divergence of the RS expansion, whereas large-order numerical calculations indicate that RS converges for this system.[1] Nevertheless, convergence properties of various intermolecular perturbation expansions are closely related to the way in which the one-electron part of the interaction is taken into account; we will return to this observation in Sect. 3 while discussing the regularized SAPT theory.

2.2
Symmetry-forcing Technique

As we have already stated, the zero-order wave function ϕ_0 and the exact wave function ψ exhibit different symmetry with respect to interchanges of electrons. Whereas ψ is fully antisymmetric for all electron permutations, ϕ_0 is only antisymmetric with respect to interchanges of electrons belonging to the same monomer (A or B). Therefore, one can improve convergence of the polarization expansion by forcing the full antisymmetry, i.e. by inserting appropriate projection operators into the perturbation equations, Eqs. 8 and 9. The specific form of these projectors depends on the specific choice of the spaces $\mathcal{H}_A$ and $\mathcal{H}_B$. In the following discussion we shall use the so-called *spin-free approach* [83, 84] in which $\mathcal{H}_X$ (X = A, B) is spanned by spatial-only functions corresponding to a specific irreducible representation (irrep) $[\lambda]$ of the symmetric group S_{N_X}, where N_X is the number of electrons in the monomer X. The choice of $[\lambda]$ determines the spin multiplicity of the monomer X since a spatial wave function of symmetry $[\lambda]$ can only be combined with a spin wave function of the conjugate symmetry $[\lambda]^\dagger$ (the Young diagrams of $[\lambda]$ and $[\lambda]^\dagger$ are a transpose of each other) to form an antisymmetric total wave function [84]. In this approach, the symmetry operators to be inserted into Eqs. 8 and 9 are the Young projectors $\mathcal{A}^{[\lambda]}$ onto the subspace of appropriate symmetry $[\lambda]$ with respect to the S_N group, $N = N_A + N_B$. $[\lambda]$

---

[1] Korona T, unpublished

must be the symmetry of one of the subspaces into which the whole space $\mathcal{H}_A \otimes \mathcal{H}_B$ decomposes under the action of S_N, cf. Eq. 9 of [57]. If the conventional, spin approach were used, and the spaces $\mathcal{H}_A$ and $\mathcal{H}_B$ were spanned by Slater determinants, the symmetry operators would be products of the antisymmetrizer and a suitable spin projection.

Most of the existing SAPT formulations can be obtained using the general symmetry-forcing technique developed in [67] and [40]. The iterative scheme of Eqs. 5 and 6 is generalized as follows,

$$\psi_n = \phi_0 + R_0(\mathcal{E}_n - V)\mathcal{F}\psi_{n-1} , \tag{10}$$

$$\mathcal{E}_n = \frac{\left\langle \phi_0 | V \mathcal{G} \psi_{n-1} \right\rangle}{\left\langle \phi_0 | \mathcal{G} \psi_{n-1} \right\rangle} , \tag{11}$$

where $\mathcal{F}$ and $\mathcal{G}$ are symmetry-forcing operators. The denominator in the energy is necessary since $\mathcal{G}$ does not have to conserve the intermediate normalization of ψ_{n-1}. Different choices of the operators $\mathcal{F}$ and $\mathcal{G}$, as well as of the function ψ_0 used to initiate the iterations, lead to different SAPT expansions listed in Table 1. In this table, as well as throughout the rest of this review, the symmetry index $[\lambda]$ will be omitted as long as it does not lead to any ambiguities.

Obviously, the RS method is obtained from the iterative process expression defined by Eq. 11 with no symmetrization performed. The simplest SAPT theory with symmetrization, taking into account the exchange part of the interaction energy in finite order, is the SRS expansion formulated in [40]. In the SRS method, the wave function corrections are taken directly from the RS theory, $\psi_{SRS}^{(n)} \equiv \psi_{RS}^{(n)}$, and the perturbation energies are calculated from the formula

$$E_{SRS}^{(n)} = N_0 \left[\left\langle \phi_0 | V \mathcal{A} \psi_{RS}^{(n-1)} \right\rangle - \sum_{k=1}^{n-1} E_{SRS}^{(k)} \left\langle \phi_0 | \mathcal{A} \psi_{RS}^{(n-k)} \right\rangle \right] , \tag{12}$$

where $N_0 = \left\langle \phi_0 | \mathcal{A} \phi_0 \right\rangle^{-1}$. It is the SRS theory that has been implemented in the general-utility closed-shell SAPT program [24] and widely applied in practical calculations [20].

Similar to SRS, but a little more complicated, is the MSMA theory introduced in [85] and [86]. The iterative process resulting from the original formulation of the MSMA theory starts from the symmetrized and intermediately normalized zero-order function $N_0 \mathcal{A} \phi_0$, and no further symmetrization is applied ($\mathcal{F} = \mathcal{G} = 1$). It has been shown in [67] that the choice $\mathcal{F} = 1$, $\mathcal{G} = \mathcal{A}$, and $\psi_0 = \phi_0$ leads to the same energy corrections $E_{MSMA}^{(n)}$, although the wave function corrections are different. The MSMA energies $E_{MSMA}^{(n)}$ and wave functions $\psi_{MSMA}^{(n)}$ can therefore be calculated from Eqs. 12 and 9, respectively, with all the RS and SRS quantities replaced by their MSMA counterparts. The

Table 1 Symmetry-adapted perturbation theories obtained using the symmetry-forcing technique. The last two columns give equation numbers from which the energy and wave function corrections of a given method can be calculated

Method	$\mathcal{F}$	$\mathcal{G}$	ψ_0	Eq. for $E^{(n)}$	Eq. for $\psi^{(n)}$
RS	1	1	ϕ_0	8	9
SRS	1	$1(\mathcal{A})^a$	ϕ_0	12	9
MSMA(a)	1	1	$N_0\mathcal{A}\phi_0$		
MSMA(b)	1	$\mathcal{A}$	ϕ_0	12	9
ELHAV	$\mathcal{A}$	$\mathcal{A}$	$N_0\mathcal{A}\phi_0$	15	13–14
JK	$\mathcal{A}$	1	$N_0\mathcal{A}\phi_0$	8	13–14
JK-1	$\mathcal{A}$	1	$N_1\mathcal{A}(\phi_0 - R_0 V\phi_0)$	8	17

[a] In the SRS theory the wave function corrections are taken from the RS method, and the energy corrections are calculated from Eq. 12.

symmetry forcing employed in the SRS and MSMA methods is commonly referred to as *weak* [67], since no symmetrization is applied in Eq. 10 defining corrections to the wave function. The theories with weak symmetry forcing cannot be expected to converge in the presence of a Pauli-forbidden continuum.

Among the theories employing the so-called *strong symmetry forcing*, for which the symmetry projector $\mathcal{A}$ appears in Eq. 10, are the ELHAV [22, 59, 60] and JK [67] methods, as well as the AM expansion [62] which does not fit into the general scheme of Eqs. 10 and 11 and will be discussed separately. The ELHAV theory has quite a few apparently different but fully equivalent formulations. Here we have chosen the one introduced in [40]. Within this approach, which can be referred to as involving the strong symmetry-forcing procedure, Eqs. 10 and 11 are iterated with $\mathcal{F} = \mathcal{G} = \mathcal{A}$ and the starting function $\psi_0 = N_0\mathcal{A}\phi_0$. The corrections $\psi_{\mathrm{ELHAV}}^{(n)}$ are then defined by the formulae

$$\psi_{\mathrm{ELHAV}}^{(1)} = N_0 R_0 \mathcal{A}(E_{\mathrm{ELHAV}}^{(1)} - V)\phi_0 \,, \tag{13}$$

$$\psi_{\mathrm{ELHAV}}^{(n)} = -R_0 V \mathcal{A}\psi_{\mathrm{ELHAV}}^{(n-1)} + \sum_{k=1}^{n} E_{\mathrm{ELHAV}}^{(k)} R_0 \mathcal{A}\psi_{\mathrm{ELHAV}}^{(n-k)} \tag{14}$$

for $n \geq 2$. In these equations, $\psi_{\mathrm{ELHAV}}^{(0)} = N_0\mathcal{A}\phi_0$, and the energies $E_{\mathrm{ELHAV}}^{(n)}$ are calculated from the formula

$$E_{\mathrm{ELHAV}}^{(n)} = \left\langle \phi_0 | V \mathcal{A}\psi_{\mathrm{ELHAV}}^{(n-1)} \right\rangle - \sum_{k=1}^{n-1} E_{\mathrm{ELHAV}}^{(k)}\left\langle \phi_0 | \mathcal{A}\psi_{\mathrm{ELHAV}}^{(n-k)} \right\rangle. \tag{15}$$

Since the unphysical components of $\psi_{\mathrm{ELHAV}}^{(n)}$ are explicitly projected out on the r.h.s. of Eqs. 13 and 14, the ELHAV method may converge despite the presence of a continuum. However, unlike the methods employing weak symmetry forcing, its second-order energy $E_{\text{ELHAV}}^{(2)}$ does not recover correctly the leading C_n/R^n term in the asymptotic expansion of the interaction energy [65]. This is a serious drawback, both from the theoretical (the phenomena of induction and dispersion are not correctly described) and practical (low-order energies are highly inaccurate for large R) point of view. The origin of the wrong asymptotics of the second-order ELHAV energy is explained in detail in [67].

To correct the asymptotics of the second-order ELHAV energy, Jeziorski and Kołos suggested a new approach [67], referred to as the JK method. This approach differs from ELHAV by the absence of the antisymmetrizer in the energy expression (Eq. 11) i.e. $\mathcal{F} = \mathcal{A}$, $\mathcal{G} = 1$, and $\psi_0 = N_0 \mathcal{A} \phi_0$. As discussed in detail below, the second-order JK energy exhibits the correct asymptotic behavior, i.e. recovers the exact value of C_6 and a few higher van der Waals constants. The third- and higher-order JK energies, however, have incorrect asymptotic behavior, and the term C_{11}/R^{11} for spherically symmetric atoms, or C_9/R^9 for polar molecules, is not fully recovered by a finite-order JK expansion.

A method to further refine the low-order asymptotics of the JK theory has been proposed in [69]. In this approach, one retains the form of $\mathcal{F}$ and $\mathcal{G}$, but improves the function used to start the iterative process, using suitable functions from the RS theory. If the iterations are started from $\psi_0 = N_1 \mathcal{A}(\phi_0 + \psi_{\text{RS}}^{(1)}) = N_1 \mathcal{A}(\phi_0 - R_0 V \phi_0)$, where the constant

$$N_1 = \frac{1}{\left\langle \phi_0 | \mathcal{A} \phi_0 \right\rangle - \left\langle \phi_0 | \mathcal{A} R_0 V \phi_0 \right\rangle} \tag{16}$$

enforces the intermediate normalization of ψ_0, one obtains the so-called JK-1 expansion. The individual corrections $\psi_{\text{JK-1}}^{(n)}$ are calculated from the formula

$$\psi^{(n)} = - R_0 V \mathcal{A} \psi^{(n-1)} + \sum_{k=1}^{n} E^{(k)} R_0 \mathcal{A} \psi^{(n-k)} - N_1^{(n-1)} \mathcal{A} R_0 V \phi_0 \tag{17}$$

$$+ N_1^{(n)} \mathcal{A} \phi_0 + N_1^{(n-1)} R_0 (V - \left\langle \phi_0 | V \phi_0 \right\rangle) \mathcal{A} \phi_0 + N_1^{(n-2)} R_0 \mathcal{A} V R_0 V \phi_0$$

$$- N_1^{(n-2)} R_0 V \mathcal{A} R_0 V \phi_0 \,,$$

for $n \geq 1$, where

$$N_1^{(k)} = \begin{cases} N_0^{k+1} \left\langle \phi_0 | \mathcal{A} R_0 V \phi_0 \right\rangle^k & k \geq 0 \\ 0 & k < 0 \end{cases} \tag{18}$$

and $\psi^{(0)} = N_0 \mathcal{A} \phi_0$. The energy corrections are obtained from Eq. 8, with the RS functions replaced by the corresponding ones calculated from Eq. 17. The derivation of Eq. 17 involves the commutation relation

$$[\mathcal{E}_n - V, \mathcal{A}] = [H_0 - E_0, \mathcal{A}] \tag{19}$$

and is presented in detail in [69]. One can show that the improvement of ψ_0 resulting from the use of the first-order RS wave function makes the resulting theory asymptotically correct to one order further than JK, i.e. JK-1 exhibits the correct asymptotics in the second and third order. The asymptotic behavior of $E_{\text{JK}-1}^{(4)}$ and higher corrections remains incorrect.

It is quite instructive to look in more detail into the lowest-order energy corrections and their asymptotics for the SAPT theories discussed so far. For this purpose, we introduce a simplified notation $\langle X \rangle \equiv \langle \phi_0 | X \phi_0 \rangle$ for any operator X. Using the formulae referenced in Table 1, one can show that the low-order RS, SRS, ELHAV, JK, and JK-1 corrections are equal to

$$E_{\text{RS}}^{(1)} = \langle V \rangle \tag{20}$$

$$E_{\text{RS}}^{(2)} = - \langle V R_0 V \rangle \tag{21}$$

$$E_{\text{RS}}^{(3)} = \langle V R_0 V R_0 V \rangle - \langle V \rangle \langle V R_0^2 V \rangle \tag{22}$$

$$E_{\text{SRS}}^{(1)} = N_0 \langle V \mathcal{A} \rangle \tag{23}$$

$$E_{\text{SRS}}^{(2)} = - N_0 \langle V \mathcal{A} R_0 V \rangle + N_0^2 \langle V \mathcal{A} \rangle \langle \mathcal{A} R_0 V \rangle \tag{24}$$

$$E_{\text{SRS}}^{(3)} = N_0 \langle V \mathcal{A} R_0 V R_0 V \rangle - N_0 \langle V \rangle \langle V \mathcal{A} R_0^2 V \rangle - N_0 \langle V \mathcal{A} \rangle \langle \mathcal{A} R_0 V R_0 V \rangle \tag{25}$$
$$+ N_0 \langle V \mathcal{A} \rangle \langle V \rangle \langle \mathcal{A} R_0^2 V \rangle - N_0 \langle V \mathcal{A} R_0 V \rangle \langle \mathcal{A} R_0 V \rangle$$
$$+ N_0^2 \langle V \mathcal{A} \rangle \langle \mathcal{A} R_0 V \rangle^2$$

$$E_{\text{ELHAV}}^{(1)} = N_0 \langle V \mathcal{A} \rangle \tag{26}$$

$$E_{\text{ELHAV}}^{(2)} = 2 N_0^2 \langle V \mathcal{A} R_0 \mathcal{A} \rangle \langle V \mathcal{A} \rangle - N_0 \langle V \mathcal{A} R_0 \mathcal{A} V \rangle - N_0^3 \langle V \mathcal{A} \rangle^2 \langle \mathcal{A} R_0 \mathcal{A} \rangle \tag{27}$$

$$E_{\text{JK}}^{(1)} = N_0 \langle V \mathcal{A} \rangle \tag{28}$$

$$E_{\text{JK}}^{(2)} = - N_0 \langle V \mathcal{A} R_0 V \rangle + N_0^2 \langle V \mathcal{A} \rangle \langle \mathcal{A} R_0 V \rangle \tag{29}$$

$$E_{\text{JK}}^{(3)} = - N_0^2 \langle V \mathcal{A} \rangle \langle V R_0 V \mathcal{A} R_0 \mathcal{A} \rangle + N_0 \langle V R_0 V \mathcal{A} R_0 \mathcal{A} V \rangle \tag{30}$$
$$+ N_0^3 \langle V \mathcal{A} \rangle^2 \langle V R_0 \mathcal{A} R_0 \mathcal{A} \rangle - N_0^2 \langle V \mathcal{A} \rangle \langle V R_0 \mathcal{A} R_0 \mathcal{A} V \rangle$$
$$+ N_0^3 \langle V \mathcal{A} \rangle \langle \mathcal{A} R_0 V \rangle^2 - N_0^2 \langle \mathcal{A} R_0 V \rangle \langle V \mathcal{A} R_0 V \rangle$$

$$E_{JK-1}^{(1)} = N_0 \langle V\mathcal{A} \rangle \tag{31}$$

$$E_{JK-1}^{(2)} = - N_0 \langle V\mathcal{A}R_0 V \rangle + 2N_0^2 \langle V\mathcal{A} \rangle \langle \mathcal{A}R_0 V \rangle - N_0 \langle V \rangle \langle \mathcal{A}R_0 V \rangle \tag{32}$$

$$E_{JK-1}^{(3)} = N_0 \langle VR_0 V\mathcal{A}R_0 V \rangle - N_0^2 \langle V\mathcal{A} \rangle \langle VR_0 \mathcal{A}R_0 V \rangle$$
$$- N_0^2 \langle V\mathcal{A} \rangle \langle VR_0 V\mathcal{A}R_0 \mathcal{A} \rangle + N_0 \langle V \rangle \langle VR_0 V\mathcal{A}R_0 \mathcal{A} \rangle$$
$$+ N_0^3 \langle V\mathcal{A} \rangle^2 \langle VR_0 \mathcal{A}R_0 \mathcal{A} \rangle - N_0^2 \langle V \rangle \langle V\mathcal{A} \rangle \langle VR_0 \mathcal{A}R_0 \mathcal{A} \rangle$$
$$+ 4N_0^3 \langle V\mathcal{A} \rangle \langle \mathcal{A}R_0 V \rangle^2 - 2N_0^2 \langle V \rangle \langle \mathcal{A}R_0 V \rangle^2$$
$$- 2N_0^2 \langle \mathcal{A}R_0 V \rangle \langle V\mathcal{A}R_0 V \rangle \tag{33}$$

It has been shown [66] that the RS expansion recovers the exact R^{-n} asymptotics of the interaction energy,

$$\left| \mathcal{E} - \sum_{n=1}^{N} E_{RS}^{(n)} \right| = O(R^{-3N-3}) . \tag{34}$$

Therefore, we may analyze the asymptotic behavior of various SAPT formalisms by comparing the energy corrections to the RS ones. It may be shown that the insertion of a *single* symmetry projector $\mathcal{A}$ into an expectation value expression $\langle \cdots \rangle$, containing any multiple product of the operators R_0 and V, and a simultaneous multiplication of the whole expression by N_0, does not influence the R^{-n} asymptotic behavior. In the simplest example, the expressions $\langle V \rangle$ and $N_0 \langle V\mathcal{A} \rangle$ exhibit the same asymptotics, i.e. the first-order SRS exchange energy $E_{SRS}^{(1)} - E_{RS}^{(1)} = N_0 \langle V\mathcal{A} \rangle - \langle V \rangle$ vanishes exponentially with R. However, no similar asymptotic equality exists for expressions $\langle \cdots \rangle$ involving more than one symmetry projector. For example, the asymptotics of the expression $\langle V\mathcal{A}R_0 \mathcal{A}V \rangle$, entering the second-order ELHAV energy, is determined not only by $\langle VR_0 V \rangle$, but contains also the so-called "double exchange" terms of the form $\langle V\mathcal{P}_{ij}R_0\mathcal{P}_{ij}V \rangle$, where $\mathcal{P}_{ij}$ transposes the coordinates of electrons i and j, which do not vanish exponentially with R [67].

By removing the single occurrences of $\mathcal{A}$, as explained in the previous paragraph, one sees that the second- and third-order SRS energies are asymptotically equivalent to the corresponding RS ones. To derive this result we used the fact that $R_0\phi_0 = 0$, therefore, e.g. $\langle R_0 V \rangle = 0$, and thus the quantity $\langle \mathcal{A}R_0 V \rangle$ vanishes exponentially with R. In fact, any SRS energy correction $E_{SRS}^{(n)}$ exhibits the same correct asymptotic behavior as $E_{RS}^{(n)}$ [66]. On the other hand, the formula for the second-order ELHAV energy contains expressions $\langle \cdots \rangle$ involving double antisymmetrizers, and the asymptotic behavior of such expressions is by no means related to that of $E_{RS}^{(2)}$. For the JK theory, the second-order energy is the same as in the SRS approach, so the asymptotic behavior is correct. However, $E_{JK}^{(3)}$ contains expressions $\langle \cdots \rangle$ with double antisymmetrizers, and it does not behave correctly in the asymptotic limit.

In the JK-1 theory developed in [69], the second-order energy differs from the corresponding SRS value. However, by removing the single occurrences of $\mathcal{A}$ one may easily show that this difference vanishes exponentially, so the asymptotic behavior of $E_{JK-1}^{(2)}$ is also correct. The same result holds for the third order. It is worth noting that $E_{JK-1}^{(3)}$ is asymptotically correct despite the fact that the 3rd through 6th terms in Eq. 33 individually exhibit unphysical long-range $(1/R^9)$ behavior. It turns out that the long-range parts of these terms mutually cancel out and the sum of these terms becomes proportional to $N_0 \langle V\mathcal{A} \rangle - \langle V \rangle$, an exponentially vanishing quantity. This cancellation does not take place in $E_{JK-1}^{(4)}$ and in higher corrections, and no non-regularized theory is both convergent and asymptotically correct to any order of perturbation theory.

Obviously, one can improve the function ψ_0 used to start the iterative process, Eqs. 10–11, further, using higher wave functions of the polarization theory, e.g. setting $\psi_0 = N_2\mathcal{A}(\phi_0 + \psi_{RS}^{(1)} + \psi_{RS}^{(2)})$, where the constant N_2 is such that ψ_0 fulfills the intermediate normalization condition. The JK-2 method obtained with this starting point should be asymptotically correct to the fourth order of perturbation theory. However, this approach is not suitable for the construction of a theory that can converge in the presence of the Pauli-forbidden continuum and is at the same time asymptotically correct to any order. For this purpose, one must use the regularization technique described in Sect. 3.

High-order convergence behavior of several other SAPT expansions, including the Murrell-Shaw-Musher-Amos (MSMA) [85, 86], ELHAV, and JK theories, was studied in [69] on the same example of the LiH system. It was shown that expansions such as ELHAV and JK converge despite the presence of the Pauli-forbidden continuum, and that the asymptotic correctness of the second-order JK energy leads to much better low-order convergence compared to ELHAV. It turned out, however, that in practical applications the asymptotic behavior of third- and higher-order corrections is far less significant than the asymptotics of the second-order energy, and JK-1 provides no improvement over JK except for very large intermonomer distances.

2.3
Conjugate Formulation of SAPT and the HS Theory

In view of the fact that the operators V and $\mathcal{A}$ do not commute, the order of operators chosen in the definition of the symmetry-forcing technique (Eqs. 10 and 11) is not the only possible one and a different theory is obtained if one replaces the operator $V\mathcal{A}$ by its Hermitian conjugate, the operator $\mathcal{A}V$. This conjugate formulation of SAPT allows one to define the Amos-Musher perturbation theory [62]. It was also employed in one of the original formulations of the ELHAV method [59].

The Amos-Musher theory in its pure, original form [62] is simply the RS perturbation theory with the zero-order Hamiltonian H_0 and the perturbation equal to $\mathcal{A}V$. Thus, the energy and wave function corrections $E_{AM}^{(n)}$ and $\psi_{AM}^{(n)}$ are obtained from Eqs. 8 and 9, respectively, with all the occurrences of V replaced by $\mathcal{A}V$. As in the definition of the symmetrized Rayleigh-Schrödinger approach, one may employ the functions $\psi_{AM}^{(n)}$ in the SRS energy expression, Eq. 12, to define another set of energy corrections, which will be referred to as the symmetrized AM (SAM) energies. It turns out that the low-order SAM corrections are much more accurate than the corresponding pure AM values [72, 87].

Another modification of the original Amos-Musher theory has been proposed by Adams [55]. The modified AM Hamiltonian takes the form

$$H_{AM} = H_0 + \mathcal{A}(V - D) \tag{35}$$

where D is a (constant) offset parameter chosen such that the perturbation expansion in powers of $\mathcal{A}(V - D)$ converges faster than the original one. Common choices for D are the first-order polarization energy $\langle \phi_0 | V \phi_0 \rangle$ and the Heitler-London energy $N_0 \langle \phi_0 | V \mathcal{A} \phi_0 \rangle$. The interaction energy $\mathcal{E}$ can be obtained by adding D to the sum of the AM corrections or by summing up the SAM energies (no addition of D is needed in this case). The choice of D has some impact on the convergence properties of the AM method but is practically inconsequential when the SAM expansion is used; the relevant numerical data for the lithium hydride and three choices of D are given in Table 2 and Fig. 1 for the triplet and singlet states, respectively. A similar result has been obtained for a triplet He atom interacting with a ground-state H atom [87]. In all cases, the low-order SAM results are much more accurate than the ones obtained with the nonsymmetrized AM approach.

To obtain the corrections of the ELHAV theory [59] in the conjugate approach, one starts from the equation

$$(H_0 - E_0)\psi + \mathcal{A}(V - \mathcal{E})\psi = 0, \tag{36}$$

substitutes $V \rightarrow \zeta V$, and expands ψ and $\mathcal{E}$ in powers of ζ. The relationship between the approach of Eq. 36, the formulation by van der Avoird [60], and the symmetry-forcing derivation of the ELHAV theory shown in Sect. 2.2 has been discussed in detail in [40]. All three approaches lead to identical energy corrections; however, the wave function corrections are different.

It is worth mentioning that the development of perturbation theories from the equation

$$(H_0 - E_0)\tilde{\psi} + (V - \mathcal{E})\mathcal{A}\tilde{\psi} = 0 \tag{37}$$

Table 2 Convergence of the conventional AM and SAM expansions for the triplet state of LiH, for $R = 11.5$ bohr and three different values of the offset parameter: $D = 0$ (columns marked "0"), $D = \langle \phi_0 | V \phi_0 \rangle$ (columns marked "pol"), and $D = N_0 \langle \phi_0 | V \mathcal{A} \phi_0 \rangle$ (columns marked "HL"). The numbers listed are percent errors with respect to the FCI interaction energy

		AM			SAM	
n	0	pol	HL	0	pol	HL
2	−118.0469	−109.5340	−160.1655	−80.7499	−80.7501	−80.7488
3	−85.8639	−80.5432	−112.1885	−32.2093	−32.2100	−32.2062
4	−58.5816	−55.2563	−75.0332	−13.0982	−13.0989	−13.0945
5	−38.6454	−36.5674	−48.9257	−5.4104	−5.4110	−5.4071
6	−25.0044	−23.7060	−31.4279	−2.2656	−2.2661	−2.2630
7	−15.9887	−15.1774	−20.0018	−0.9609	−0.9612	−0.9590
8	−10.1477	−9.6409	−12.6548	−0.4126	−0.4129	−0.4113
9	−6.4095	−6.0929	−7.9755	−0.1794	−0.1796	−0.1785
10	−4.0354	−3.8376	−5.0135	−0.0790	−0.0792	−0.0785
15	−0.3905	−0.3718	−0.4834	−0.0016	−0.0016	−0.0016
20	−0.0373	−0.0356	−0.0462	0.0000	0.0000	0.0000
25	−0.0036	−0.0034	−0.0044	0.0000	0.0000	0.0000
30	−0.0003	−0.0004	−0.0005	0.0000	0.0000	0.0000

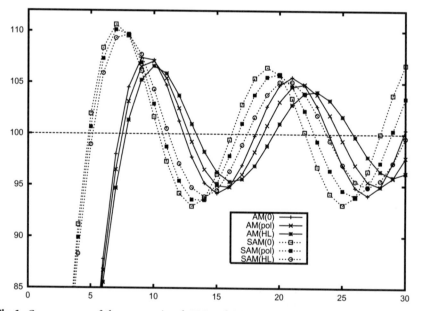

Fig. 1 Convergence of the conventional AM and SAM expansions for the singlet state of LiH, for $R = 3.015$ bohr and three different values of the offset parameter: $D = 0$ (curves marked "0"), $D = \langle \phi_0 | V \phi_0 \rangle$ (curves marked "pol"), and $D = N_0 \langle \phi_0 | V \mathcal{A} \phi_0 \rangle$ (curves marked "HL"). The numbers displayed are percentages of the FCI interaction energy recovered in the nth order perturbation treatment

(this equation is conjugate to Eq. 36 used by Hirschfelder [59]) requires some extra caution. Suppose that one substitutes $V \to \zeta V$ and performs the expansion in powers of ζ directly in Eq. 37. The first-order wave function is then given by

$$\tilde{\psi}_{\text{ELHAV}}^{(1)} = R_0(E^{(1)} - V)\mathcal{A}\phi_0 , \tag{38}$$

where $E^{(1)}$ is the first-order energy. This function does not vanish when $R \to \infty$. On the other hand, in all the perturbation theories defined before, including the ELHAV and JK methods, $\psi^{(1)}$, and all subsequent $\psi^{(n)}$, vanish identically at infinite intermonomer separations. One can easily show that the difference $\psi_{\text{ELHAV}}^{(1)} - N_0 \tilde{\psi}_{\text{ELHAV}}^{(1)}$ is equal to

$$N_0 R_0[V - E^{(1)}, \mathcal{A}]\phi_0 = - N_0 R_0[H_0 - E_0, \mathcal{A}]\phi_0 = \phi_0 - N_0 \mathcal{A}\phi_0 \tag{39}$$

and, consequently, the whole correction $\tilde{\psi}_{\text{ELHAV}}^{(1)}$ does not vanish when $R \to \infty$. One can prove by induction that the same holds for all $\tilde{\psi}_{\text{ELHAV}}^{(n)}$, $n > 1$. The same problem is encountered if one starts from Eq. 36 and does not expand it directly, but substitutes $\psi = N_0 \mathcal{A}\phi_0 + \chi$ and applies the commutation relation

$$[\mathcal{E} - V, \mathcal{A}] = [H_0 - E_0, \mathcal{A}] \tag{40}$$

to obtain an expansion for χ (exactly as was done in [69] to derive the JK-1 expansion).

The nature of the problem described above can be understood clearly when one considers the fact that a successful SAPT expansion either converges to the exact wave function of the dimer, or to a primitive function that is localized in the same way as ϕ_0 and gives the exact wave function when antisymmetrized. For these two cases, one should use something like $N_0 \mathcal{A}\phi_0$ and ϕ_0, respectively, as a zero-order function. The "wrong" theories described above try to do this the other way round – they either start from ϕ_0 and converge to $N_0 \mathcal{A}\phi_0$, or vice versa. Since ϕ_0 and $N_0 \mathcal{A}\phi_0$ are always completely different, even for $R \to \infty$, convergence of such a theory is bound to be very slow.

Whether the perturbation functions vanish asymptotically or not depends on the application of the commutation relation, Eq. 40, in deriving the perturbation series. This relation, as already noted by Hirschfelder [78], is somewhat paradoxical since a first-order quantity on the l.h.s. is equated to a (nonvanishing) zero-order quantity on the r.h.s. In other words, the equality $[\mathcal{E}(\zeta) - \zeta V, \mathcal{A}] = [H_0 - E_0, \mathcal{A}]$ is not valid for arbitrary ζ, but only for $\zeta = 1$. Thus, each application of Eq. 40 changes the way the order of perturbation theory is defined. This paradox is analyzed in more detail in a recent publication by Adams [88].

All the SAPT methods considered so far have used a single operator $\mathcal{A}^{[\lambda]}$ projecting onto a specific subspace $\mathcal{H}^{[\lambda]} \subset \mathcal{H}_A \otimes \mathcal{H}_B$. The HS perturbation theory, introduced by Hirschfelder and Silbey [50], follows a different, multistate philosophy. It performs a perturbation expansion for a primitive function

$$\Phi = \sum_{[\lambda]} c^{[\lambda]} \psi^{[\lambda]} , \tag{41}$$

where the sum goes over all asymptotically degenerate eigenstates $\psi^{[\lambda]}$ of H which dissociate into the specific states of the monomers, i.e. over all permutation symmetries $[\lambda]$ into which the Hilbert space $\mathcal{H}_A \otimes \mathcal{H}_B$ decomposes under the action of S_N, including the symmetries that lead to unphysical, Pauli-forbidden states. The primitive function in the HS method is defined uniquely by the *localization conditions* [89]

$$\left\langle \mathcal{A}^{[\lambda]} \phi_0 | (H_0 - E_0) \Phi \right\rangle = 0 \tag{42}$$

for all $[\lambda]$. Once Φ is known, the eigenstates of H can be extracted by taking projections $\mathcal{A}^{[\lambda]} \Phi$. One can easily verify that Φ satisfies the equation

$$(H_0 - E_0) \Phi = -V\Phi + \sum_{[\lambda]} \mathcal{E}^{[\lambda]} \mathcal{A}^{[\lambda]} \Phi . \tag{43}$$

Substitution $V \to \zeta V$ in Eqs. 42–43 and expansion of Φ and $\mathcal{E}^{[\lambda]}$ in powers of ζ gives the following equations for the HS perturbation corrections [49],

$$^{[\lambda]}E_{HS}^{(n)} = \left\langle \phi_0 | \mathcal{A}^{[\lambda]} \phi_0 \right\rangle^{-1} \left[\left\langle \phi_0 | V \mathcal{A}^{[\lambda]} \Phi^{(n-1)} \right\rangle \right. \tag{44}$$
$$\left. - \sum_{k=1}^{n-1} {}^{[\lambda]}E_{HS}^{(k)} \left\langle \phi_0 | \mathcal{A}^{[\lambda]} \Phi^{(n-k)} \right\rangle \right],$$

$$\left(H_0 - E_0 \right) \Phi^{(n)} = -V\Phi^{(n-1)} + \sum_{k=1}^{n} \sum_{[\lambda]} {}^{[\lambda]}E_{HS}^{(k)} \mathcal{A}^{[\lambda]} \Phi^{(n-k)} , \tag{45}$$

where $\Phi^{(0)} \equiv \phi_0$. The last term in Eq. 45 couples all asymptotically degenerate states (all symmetries $[\lambda]$). Thus, the HS theory is far more complicated in practical applications than the other approaches presented so far, especially for many-electron systems where the large number of different permutational symmetries $[\lambda]$ leads to unphysical states asymptotically degenerate with each physically allowed one.

High-order convergence studies of the HS perturbation expansion (as well as of the RS and SRS expansions) for Li − H, the simplest system including a more-than-two-electron monomer, have been presented in [57]. These studies show that the HS expansion, as the RS and SRS expansions discussed above, is indeed divergent for a system exhibiting the Pauli-forbidden continuum. However, low-order SRS and HS results turned out to be quite accurate, and it was possible to obtain extremely accurate results by summing up the perturbation corrections until these corrections start to grow in absolute value (the standard method of summing an asymptotically convergent series). Surprisingly, the HS expansion behaved no better than the much simpler SRS series, unlike in the case of interactions between one- and two-electron monomers.

3
Convergence Properties of Regularized SAPT

3.1
Methods of Regularizing the Coulomb Potential

The existence of an unphysical continuum surrounding the physical states of interacting many-electron systems is caused by the fact that, when the wave function does not obey the Pauli principle, electrons that were initially assigned to one monomer can fall into the Coulomb wells of the other monomer, ejecting some other electrons into the continuum (as in the Auger process). This is possible since the negative Coulomb wells in V are of exactly the same magnitude as the ones in H_0. The main idea behind the regularization of the Coulomb potential, a concept first employed by Herring [90] in his studies of the asymptotics of the exchange energy, is to remove all negative singularities from V so that the electrons belonging to one monomer cannot fall into the wells around the nuclei of the other monomer. Such a modified Coulomb potential should be as similar as possible to the original one. In particular, it must exhibit the same large-R asymptotics, i.e. the difference between regularized and non-regularized Coulomb potentials must vanish exponentially with the distance from the nucleus. Moreover, the one-electron integrals involving the regularized potential should be easy to evaluate.

A simple choice for a regularized Coulomb potential would be

$$v_p(r) = \begin{cases} 1/c & r \le c \\ 1/r & r > c \end{cases} \tag{46}$$

where $c > 0$ is a parameter. This is the potential used by Adams in his recent work on the regularized SAPT [71]. However, it is preferable that $v_p(r)$ is smooth for $r > 0$. Moreover, the one-electron integrals involving $v_p(r)$ of Eq. 46 and Gaussian basis functions are not so easy to compute. Therefore, two slightly more complicated analytic forms of $v_p(r)$ were employed in recent work [70]:

$$v_p(r) = \frac{1}{r}(1 - e^{-\eta r^2}), \tag{47}$$

and

$$v_p(r) = \frac{1}{r}\text{erf}(\sqrt{\omega}r),\tag{48}$$

where $\eta > 0$ and $\omega > 0$ are parameters defining the strength of the regularization, and $\text{erf}(z)$ is the standard error function

$$\text{erf}(z) = \frac{2}{\sqrt{\pi}}\int_0^z e^{-t^2}dt.\tag{49}$$

The choice defined by Eq. 47, which will be referred to as the *Gaussian regularization*, has the advantage that the matrix elements of v_p between Gaussian basis functions can be evaluated in exactly the same way as the ordinary one-electron potential energy integrals. The regularized potential of Eq. 48 was first employed by Ewald [91] in his calculations of the Madelung constants in crystals. More recently, it was used in the linear scaling electronic structure theory [92, 93] and in the description of the electron correlation cusp [94]. This potential is analytic for any r – there is no cusp at $r = 0$. It is worth noting that, since

$$\left(\frac{\omega}{\pi}\right)^{3/2}\int_{\mathbb{R}^3}\frac{e^{-\omega r'^2}}{|r-r'|}dr' = \frac{\text{erf}(\sqrt{\omega}r)}{r},\tag{50}$$

the potential of Eq. 48 can be interpreted as the electrostatic potential of a smeared unit charge, with the charge distribution defined by a Gaussian function $e^{-\omega r^2}$. Thus, the regularization defined by Eq. 48 corresponds to replacing a point nuclear charge by a smeared charge of the same total value, and it will be referred to as the smeared nuclear charge (SNC) regularization.

The difference between the original and regularized Coulomb potentials,

$$v_t(r) = \frac{1}{r} - v_p(r),\tag{51}$$

which will be referred to as the singular or residual part of the Coulomb potential, is a short-range function with a singularity at $r = 0$. We have been using the subscripts p and t to remind the reader that the potentials v_p and v_t are responsible for the polarization and tunneling aspects, respectively, of the interaction phenomenon.

The interaction operator V of two atoms A and B can now be split into its regular part V_p and singular part V_t as follows,

$$V_p = \frac{Z_A Z_B}{r_{AB}} - \sum_{i=1}^{Z_A} Z_B v_p(r_{Bi}) - \sum_{j=1}^{Z_B} Z_A v_p(r_{Aj}) + \sum_{i=1}^{Z_A}\sum_{j=1}^{Z_B}\frac{1}{r_{ij}},\tag{52}$$

and

$$V_t = - \sum_{i=1}^{Z_A} Z_B v_t(r_{Bi}) - \sum_{j=1}^{Z_B} Z_A v_t(r_{Aj}) , \tag{53}$$

where $r_{pq} = |r_p - r_q|$ denotes the distance between particles p and q. The particles in Eqs. 52–53 are the nuclei A and B (with atomic numbers Z_A and Z_B, respectively), and the electrons initially assigned to A (enumerated by i) and to B (enumerated by j). In Eqs. 52–53, as well as throughout the whole text, atomic units are used. Note that only the one-electron, attractive part of the Coulomb potential has been regularized in Eq. 52, therefore, the approach of Eqs. 52–53 will be referred to as the *one-electron regularization*. The *full regularization*, corresponding to the partitioning of V into V_p^{full} and V_t^{full} operators given by,

$$V_p^{full} = \frac{Z_A Z_B}{r_{AB}} - \sum_{i=1}^{Z_A} Z_B v_p(r_{Bi}) - \sum_{j=1}^{Z_B} Z_A v_p(r_{Aj}) + \sum_{i=1}^{Z_A} \sum_{j=1}^{Z_B} v_p(r_{ij}) , \tag{54}$$

$$V_t^{full} = - \sum_{i=1}^{Z_A} Z_B v_t(r_{Bi}) - \sum_{j=1}^{Z_B} Z_A v_t(r_{Aj}) + \sum_{i=1}^{Z_A} \sum_{j=1}^{Z_B} v_t(r_{ij}) \tag{55}$$

has also been tested but only in the case of two interacting hydrogen atoms [70]. This approach turned out to perform very similarly to the one-electron regularization [70]. It is, however, significantly more complicated computationally, since the regularized two-electron integrals are required in this case. The one-electron regularization is preferable also on the theoretical grounds since it does not affect the dispersion interaction at all. This results from the fact that the dispersion part of the interaction energy $\mathcal{E}$ does not depend on the one-electron part of V.

3.2
Regularized SRS Expansion

When one neglects the operator V_t completely, the Schrödinger equation takes the form

$$\left[H_0 + V_p(\eta) - E_0 \right] \psi_p(\eta) = \mathcal{E}_p(\eta) \psi_p(\eta) , \tag{56}$$

where it has been explicitly stated that the eigenfunction ψ_p and the eigenvalue $\mathcal{E}_p$ depend on the value of the regularization parameter η (or ω, when the SNC regularization is employed). Equation 56 can be solved by means of the standard RS perturbation theory; the resulting expansion in powers of ζ, which will be referred to as the regularized RS (R-RS) expansion, takes the form

$$\mathcal{E}_{\mathrm{p}}(\eta) = \sum_{n=1}^{\infty} E_{\mathrm{R\text{-}RS}}^{(n)}(\eta)\zeta^n, \tag{57}$$

$$\psi_{\mathrm{p}}(\eta) = \sum_{n=0}^{\infty} \psi_{\mathrm{R\text{-}RS}}^{(n)}(\eta)\zeta^n, \tag{58}$$

with the individual terms given by Eqs. 8 and 9 with V replaced by V_{p}, and the RS functions and energies replaced by their regularized counterparts (obviously, $\psi_{\mathrm{R\text{-}RS}}^{(0)} \equiv \phi_0$). Note that in Eq. 56, unlike in the non-regularized Schrödinger equation, the permutational symmetry is broken. Thus, one may expect that $\psi_{\mathrm{p}}(\eta)$ will be localized in the same way as ϕ_0. In fact, for some range of values of the regularization parameter, the function $\psi_{\mathrm{p}}(\eta)$ should be close to the exact primitive function, i.e. the function $\mathcal{A}\psi_{\mathrm{p}}(\eta)$ should provide a good approximation to the exact eigenfunction ψ for any permutational symmetry forced by the projector $\mathcal{A}$.

Knowing $\psi_{\mathrm{p}}(\eta)$, one can obtain an approximation to the exact interaction energy $\mathcal{E}$ by an SRS-like energy formula,

$$\mathcal{E} \approx \mathcal{E}_{\mathrm{R\text{-}SRS}}(\eta) = \frac{\left\langle \phi_0 | V \mathcal{A} \psi_{\mathrm{p}}(\eta) \right\rangle}{\left\langle \phi_0 | \mathcal{A} \psi_{\mathrm{p}}(\eta) \right\rangle}. \tag{59}$$

Substituting $V \to \zeta V$ and expanding Eq. 59 in powers of ζ leads to the expansion

$$\mathcal{E}_{\mathrm{R\text{-}SRS}} = \sum_{n=1}^{\infty} E_{\mathrm{R\text{-}SRS}}^{(n)}\zeta^n, \tag{60}$$

where the coefficients $E_{\mathrm{R\text{-}SRS}}^{(n)}$, which will be referred to as the regularized SRS (R-SRS) corrections, are given by Eq. 12 with $\psi_{\mathrm{RS}}^{(k)}$ replaced by $\psi_{\mathrm{R\text{-}RS}}^{(k)}$ and $E_{\mathrm{SRS}}^{(k)}$ replaced by $E_{\mathrm{R\text{-}SRS}}^{(k)}$. All the energies in Eq. 60 depend on the value of the regularization parameter η (or ω). This dependence will not be explicitly shown as long as it does not lead to ambiguities. Note that Eq. 59, as well as the expression for the corrections $E_{\mathrm{R\text{-}SRS}}^{(n)}$, contains the full interaction operator V, not just V_{p}. In the limit $\eta \to \infty$ (for the Gaussian regularization) or $\omega \to \infty$ (for the SNC regularization) the R-RS and R-SRS expansions defined above are identical to the ordinary RS and SRS theories, respectively.

For a suitable range of values of the regularization parameter, the R-SRS expansion, unlike the non-regularized SRS theory, can be expected to converge since the neglect of V_{t} shifts the unphysical continuum upwards in the energy, possibly above the physical states. Moreover, as $v_{\mathrm{t}}(r)$ is a short-range potential, the R-RS and R-SRS energy corrections exhibit the same correct asymptotic behavior as the standard RS and SRS theories. The main drawback of the R-SRS series is that its sum $\mathcal{E}_{\text{R-SRS}}(\eta)$ differs somewhat from the exact interaction energy $\mathcal{E}$. To account for this difference, one has to construct a theory that takes into account not only V_p, but also V_t. There are several possible choices of such a theory, as discussed in the next subsections.

3.3
A Posteriori Inclusion of V_t

As a first step towards construction of a regularized SAPT theory that includes both V_p and V_t, let us note that if ψ_p and $\mathcal{E}_p$ are known, the remaining part of the interaction energy can be recovered by means of a perturbation expansion in powers of V_t. Since this expansion does not influence the, already correct, asymptotics of $\mathcal{E}_p$, we can now employ a method that is convergent despite the presence of the Pauli-forbidden continuum, i.e. the EL-HAV, AM, or JK theory. Unlike the case of non-regularized expansions, there is no asymptotics-related reason to expect that the JK method will perform better than the other two. If we choose the ELHAV theory, the successive corrections to the energy and the wave function, referred to as the regularized ELHAV (R-ELHAV) corrections, are obtained from the formulae [40]

$$E_{\text{R-ELHAV}}^{(n)} = \left\langle \psi_p | V_t \mathcal{A} \psi_{\text{R-ELHAV}}^{(n-1)} \right\rangle - \sum_{k=1}^{n-1} E_{\text{R-ELHAV}}^{(k)} \left\langle \psi_p | \mathcal{A} \psi_{\text{R-ELHAV}}^{(n-k)} \right\rangle, \qquad (61)$$

$$\psi_{\text{R-ELHAV}}^{(0)} \equiv N_p \mathcal{A} \psi_p, \qquad (62)$$

$$\psi_{\text{R-ELHAV}}^{(1)} = N_p \mathcal{R}_p \mathcal{A} \left(E_{\text{R-ELHAV}}^{(1)} - V_t \right) \psi_p, \qquad (63)$$

and

$$\psi_{\text{R-ELHAV}}^{(n)} = -\mathcal{R}_p V_t \mathcal{A} \psi_{\text{R-ELHAV}}^{(n-1)} + \sum_{k=1}^{n} E_{\text{R-ELHAV}}^{(k)} \mathcal{R}_p \mathcal{A} \psi_{\text{R-ELHAV}}^{(n-k)} \qquad (64)$$

for $n \geq 2$. In these equations, $\mathcal{R}_p$ is the ground-state reduced resolvent of the operator $H_0 + V_p$, and $N_p = \langle \psi_p | \mathcal{A} \psi_p \rangle^{-1}$. If one chose to use the JK method instead of ELHAV, the wave function corrections $\psi_{\text{R-JK}}^{(n)}$ would be calculated from Eqs. 62–64 as well, only the energy corrections $E_{\text{R-JK}}^{(n)}$ would be defined differently,

$$E_{\text{R-JK}}^{(n)} = \left\langle \psi_p | V_t \psi_{\text{R-JK}}^{(n-1)} \right\rangle. \qquad (65)$$

Once the wave function corrections $\psi_{\text{R-ELHAV}}^{(k)}$ have been calculated, one can use the SRS-like formula, Eq. 59, and define an alternative expansion for the interaction energy, called in [70] the R2-ELHAV expansion. The expression for the R2-ELHAV energy corrections is

$$
E^{(n)}_{\text{R2-ELHAV}} = N_{0\text{p}} \Bigg[\left\langle \phi_0 \middle| V \mathcal{A} \psi^{(n-1)}_{\text{R-ELHAV}} \right\rangle \tag{66}
$$

$$
- \sum_{k=1}^{n-1} E^{(k)}_{\text{R2-ELHAV}} \left\langle \phi_0 \middle| \mathcal{A} \psi^{(n-k)}_{\text{R-ELHAV}} \right\rangle \Bigg],
$$

where

$$
N_{0\text{p}} = \frac{\left\langle \psi_{\text{p}} \middle| \mathcal{A} \psi_{\text{p}} \right\rangle}{\left\langle \phi_0 \middle| \mathcal{A} \psi_{\text{p}} \right\rangle} . \tag{67}
$$

The R2-ELHAV approach has the slight advantage that $E^{(1)}_{\text{R2-ELHAV}} = \mathcal{E}_{\text{R-SRS}}$, so the second- and higher-order corrections can be viewed as small contributions improving the, already quite accurate, infinite-order R-SRS energy.

3.4
Double Perturbation Approach

If the R-ELHAV expansion is able to effectively reproduce the part of the interaction energy missing in $\mathcal{E}_{\text{R-SRS}}$ in a low-order treatment (as will be seen in Sect. 4, this is the case), it is desirable to extend this theory to obtain a perturbation expansion that starts from ϕ_0 and takes both V_{p} and V_{t} into account. For this purpose, the most straightforward idea is to develop some double perturbation expansion in V_{p} and V_{t} which treats these two perturbations in an SRS-like and ELHAV-like way, respectively. The formulae defining the wave function corrections in this double perturbation theory can be obtained by expanding the equation

$$
(H_0 - E_0 + \mu V_{\text{p}})\psi + \mathcal{A}(\nu V_{\text{t}} - \mathcal{E})\psi = 0 \tag{68}
$$

in powers of μ and ν. Note that if $V_{\text{p}} = 0$ and $V_{\text{t}} = V$, Eq. 68 defines the ELHAV theory in Hirschfelder's formulation [59]. Assuming the convention that the first index refers to the order in the perturbation V_{p}, and setting $\psi^{(0,0)} \equiv \phi_0$, one obtains

$$
\psi^{(i,j)} = - \mathcal{R}_0 V_{\text{p}} \psi^{(i-1,j)} - \mathcal{R}_0 \mathcal{A} V_{\text{t}} \psi^{(i,j-1)} + \mathcal{R}_0 \mathcal{A} \sum_{k=0}^{i} \sum_{l=0}^{j}{}' E^{(k,l)} \psi^{(i-k,j-l)} ,
$$

$$
\tag{69}
$$

where the energies $E^{(i,j)}$ are given by

$$E^{(i,j)} = N_0 \left[\left\langle \phi_0 | V_p \psi^{(i-1,j)} \right\rangle + \left\langle \phi_0 | \mathcal{A} V_t \psi^{(i,j-1)} \right\rangle \right. \tag{70}$$

$$\left. - \sum_{k=0}^{i} \sum_{l=0}^{j}{}'' E^{(k,l)} \left\langle \phi_0 | \mathcal{A} \psi^{(i-k,j-l)} \right\rangle \right],$$

the prime in the summation over (k, l) denotes omission of the term $k = l = 0$, and the double prime – omission of the terms $k = l = 0$ and $k = i$, $l = j$. To keep Eqs. 69 and 70 compact, we defined here $\psi^{(i,j)} \equiv 0$ whenever $i < 0$ or $j < 0$. An alternative formula for the energy corrections, corresponding to the R2-ELHAV approach, can be obtained by expanding the equation

$$\mathcal{E} = \frac{\left\langle \phi_0 | (\mu V_p + \nu V_t) \mathcal{A} \psi \right\rangle}{\left\langle \phi_0 | \mathcal{A} \psi \right\rangle} \tag{71}$$

in powers of μ and ν. The result is

$$\mathcal{E}^{(i,j)} = N_0 \left[\left\langle \phi_0 | V_p \mathcal{A} \psi^{(i-1,j)} \right\rangle + \left\langle \phi_0 | V_t \mathcal{A} \psi^{(i,j-1)} \right\rangle \right. \tag{72}$$

$$\left. - \sum_{k=0}^{i} \sum_{l=0}^{j}{}'' \mathcal{E}^{(k,l)} \left\langle \phi_0 | \mathcal{A} \psi^{(i-k,j-l)} \right\rangle \right].$$

Double perturbation theory calculations are very time-consuming if one wants to go to high orders. Therefore, it would be highly advantageous to combine V_p and V_t in a single perturbation theory, related to Eqs. 69–72 as closely as possible. We will present such a theory – the R-SRS+ELHAV method [72] – in the next subsection.

3.5
The "All-in-one" R-SRS+ELHAV Theory

To derive perturbation equations for a theory that uses ϕ_0 as the zero-order function, takes into account both V_p and V_t, and avoids the complications of a double perturbation theory framework, we start from the following equation [72],

$$\left[H_0 - E_0 + V_p - \mathcal{E}_p + \mathcal{A}(V_t - \mathcal{E}_t) \right] \psi = 0, \tag{73}$$

where $\mathcal{E}_t = \mathcal{E} - \mathcal{E}_p$. Performing the substitution $V_p \to \zeta V_p$, $V_t \to \zeta V_t$, and using the already known R-RS perturbation expansion for $\mathcal{E}_p$, Eq. 57, one finds that the coefficients in the expansions

$$\mathcal{E}_t(\zeta) = \sum_{n=1}^{\infty} E_t^{(n)} \zeta^n , \tag{74}$$

$$\psi(\zeta) = \sum_{n=0}^{\infty} \psi_t^{(n)} \zeta^n \tag{75}$$

can be calculated from the equations

$$E_t^{(n)} = N_0 \left[\left\langle \phi_0 | (V_p + \mathcal{A} V_t) \psi_t^{(n-1)} \right\rangle - \sum_{k=1}^{n-1} E_t^{(k)} \left\langle \phi_0 | \mathcal{A} \psi_t^{(n-k)} \right\rangle - E_{\text{R-RS}}^{(n)} \right] \tag{76}$$

and

$$\psi_t^{(n)} = - R_0 \left[(V_p + \mathcal{A} V_t) \psi_t^{(n-1)} - \sum_{k=1}^{n-1} E_{\text{R-RS}}^{(k)} \psi_t^{(n-k)} - \sum_{k=1}^{n} E_t^{(k)} \mathcal{A} \psi_t^{(n-k)} \right], \tag{77}$$

where $\psi_t^{(0)} = \phi_0$. Once the wave function corrections $\psi_t^{(n)}$ are known, the energy corrections are calculated from the SRS-like formula, Eq. 15, with $\psi_{\text{ELHAV}}^{(k)}$ replaced by $\psi_t^{(k)}$. The SAPT expansion defined in this way may be regarded as a single-step combination of the R-SRS method and the ELHAV theory and will be referred to as the R-SRS+ELHAV expansion. It has been found that the low-order R-SRS+ELHAV energies are significantly more accurate than the corresponding sums of the coefficients $E_{\text{R-RS}}^{(n)}$ and $E_t^{(n)}$ (as is also the case for the non-regularized AM and SAM energies, cf. Sect. 2.3).

One should note that in order to calculate $\psi_t^{(n)}$ one has to obtain the R-RS energy corrections up to nth order from a separate expansion. This is not a significant computational complication, and it should be contrasted with the R-ELHAV theory, where to calculate $\psi_{\text{R-ELHAV}}^{(n)}$ for any n one must know the energy $\mathcal{E}_p$ to infinite order in V_p. As in the preceding subsection, we employed the conjugate formulation of SAPT (cf. Eq. 36 of Sect. 2.3) to obtain the starting point for the development of the R-SRS+ELHAV method. One can also try to develop a theory by an extension of the symmetry-forcing formalism, i.e. starting from the equation

$$\left[H_0 - E_0 + V_p - \mathcal{E}_p + (V_t - \mathcal{E}_t) \mathcal{A} \right] \psi = 0 . \tag{78}$$

The perturbation equations for such a conjugate R-SRS+ELHAV formalism [95] turned out to be significantly more complicated than Eqs. 76–77, both formally and computationally. However, numerical results that we have obtained using this conjugate approach (for the LiH system) differed insignificantly from the results of the R-SRS+ELHAV theory discussed here.

3.6
The R-SRS+SAM Approach

The R-SRS+ELHAV theory outlined above is not the only possible way of including both V_p and V_t in a single perturbation treatment. Another method of doing so has been introduced by Adams in a recent contribution [71]. Adams called his method the corrected SRS (cSRS), however, we will use the name R-SRS+SAM to emphasize the relation of his theory to the symmetrized Amos-Musher approach.

In the R-SRS+SAM theory, the Schrödinger equation takes the form

$$\left(H_0 + V_p + \mathcal{A}(V_t - D)\right) \psi_{\text{R-SRS+AM}} = (E_0 + \mathcal{E}_{\text{R-SRS+AM}})\psi_{\text{R-SRS+AM}} . \tag{79}$$

The choice of a particular offset D does not influence the results significantly (cf. Sect. 2.3). Note that if $V_p = 0$ and $V_t = V$, Eq. 79 would be identical to the one appearing in the AM perturbation theory [62]; in other words, the Hamiltonian $H_0 + V_p + \mathcal{A}(V_t - D)$ includes V_p and V_t in an RS-like and AM-like way, respectively. The eigenproblem 79 can be solved by means of the standard RS perturbation theory. In the resulting expansion of the R-SRS+AM wave function

$$\psi_{\text{R-SRS+AM}} = \phi_0 + \sum_{n=1}^{\infty} \psi_{\text{R-SRS+AM}}^{(n)} , \tag{80}$$

the coefficients $\psi_{\text{R-SRS+AM}}^{(n)}$ can be obtained from Eqs. 8 and 9 with V replaced by $V_p + \mathcal{A}(V_t - D)$ and all the RS corrections replaced by their R-SRS+AM counterparts. The interaction energy can be calculated as

$$\mathcal{E} = D + \sum_{n=1}^{\infty} E_{\text{R-SRS+AM}}^{(n)} . \tag{81}$$

However, significantly more accurate results are obtained when one follows the SAM (or SRS) algorithm and notes that $\mathcal{A}\psi_{\text{R-SRS+AM}}$ satisfies the Schrödinger equation with the full Hamiltonian H. Thus,

$$\frac{\left\langle \phi_0 | V \mathcal{A} \psi_{\text{R-SRS+AM}} \right\rangle}{\left\langle \phi_0 | \mathcal{A} \psi_{\text{R-SRS+AM}} \right\rangle} = \mathcal{E} \tag{82}$$

and the energy corrections can be calculated, as in the SRS method, from Eq. 12 with $\psi_{\text{RS}}^{(k)}$ replaced by $\psi_{\text{R-SRS+AM}}^{(k)}$. The corrections $E_{\text{R-SRS+SAM}}^{(n)}$ obtained in this way will be named R-SRS+SAM energies, as opposed to the nonsymmetrized R-SRS+AM energies $E_{\text{R-SRS+AM}}^{(n)}$ calculated along with the R-SRS+AM wave functions using an analog of Eq. 8.

One may note that the eigenproblems 73 and 79 differ, apart from the (insignificant) presence of the offset D in the latter, only by the term $\mathcal{A}\mathcal{E}_t$

in Eq. 73 versus $\mathcal{E}_t$ in Eq. 79. Speaking more generally, these two approaches both start from ϕ_0 and apply weak symmetry forcing to V_p and strong symmetry forcing to V_t, so they both can be viewed as refinements of the regularized SRS theory. In view of these similarities, these approaches were given similar names.

3.7
Zero-order Induction Theory

Apart from the R-SRS+SAM theory itself, Adams also proposed [71] a very interesting extension to this method in which the induction effects are included already in the zeroth-order energy and wave function – the so-called zero-order induction (ZI) theory. The ZI scheme can be employed to refine the R-SRS+ELHAV approach in the same manner as Adams used it on top of the R-SRS+SAM theory; we will refer to these methods as R-SRS+ELHAV+ZI and R-SRS+SAM+ZI, respectively. Performing third-order (first-order in the wave function) calculations for the singlet state of LiH, Adams found [71] that R-SRS+SAM+ZI provides a significant improvement over SRS for distances around the chemical minimum, i.e. around 3 bohr. For larger interatomic separations, the accuracy of the third-order R-SRS+SAM+ZI energy decreased significantly, although the results were still better than the SRS ones.

The zero-order induction approach differs from its parent R–SRS+ELHAV (Eq. 73) or R–SRS+SAM (Eq. 79) theory in the specific choice of the zero-order Hamiltonian and the regular part of the perturbation operator. These operators are replaced by new operators $\widetilde{H}_0 = \widetilde{H}_A + \widetilde{H}_B$ and $\widetilde{V}_p$ defined such that the effects of the (regularized) induction interaction are included in the zeroth order. Specifically, one can set $\widetilde{H}_A = H_A + \Omega_B$ and $\widetilde{H}_B = H_B + \Omega_A$, where

$$\Omega_B = -\sum_{i\in A} Z_B v_p(r_{Bi}) + \sum_{i\in A} \int \frac{1}{r_{ij}} \rho_B^{(0)}(r_j)\mathrm{d}r_j \tag{83}$$

is the operator of the electrostatic potential of atom B resulting from the regularized Coulomb attraction of the nucleus and the repulsion of the unperturbed electronic charge distribution $\rho_B^{(0)}(r_j)$ of atom B. The definition of Ω_A is obtained by interchanging A and B in Eq. 83. In accordance with the changes in H_0, the long-range part of the perturbation now takes the form $\widetilde{V}_p = V_p - \Omega_A - \Omega_B$, and the short-range part remains unchanged, $\widetilde{V}_t = V_t$. The zero-order wave function has the form $\widetilde{\phi}_0 = \widetilde{\phi}_A \widetilde{\phi}_B$ and the zero-order energy is $\widetilde{E}_0 = \widetilde{E}_A + \widetilde{E}_B$, where $\widetilde{H}_X \widetilde{\phi}_X = \widetilde{E}_X \widetilde{\phi}_X$ for X = A, B. Note that this definition of $\widetilde{H}_0$ differs from the original Adams' formulation [71] by the absence of a small combinatorial factor multiplying V_t. This difference does not appear to be significant in practice.

The X+ZI energy corrections $E_{X+ZI}^{(n)}$, where X = R-SRS+ELHAV or X = R–SRS+SAM, are now calculated as in its parent approach, except that all functions and operators are replaced by their tilded counterparts. The interaction energy differs from the sum of the X+ZI corrections by the induction contribution $\tilde{E}_0 - E_0$ contained in the zero-order energy $\tilde{E}_0$.

It is worth noting that the R-SRS+ELHAV+ZI theory inherits all the advantages of the R-SRS+ELHAV method. The V_t perturbation is treated as in the ELHAV theory, so the R-SRS+ELHAV+ZI expansion may converge despite the presence of the Pauli-forbidden continuum. Simultaneously, the long-range perturbation V_p is treated such that the correct asymptotics of the interaction energy is ensured. The same is true for the R-SRS+SAM+ZI approach. It should also be emphasized that the ZI procedure makes sense only when the electron-nucleus attraction is regularized. Otherwise, the singular part of Ω_X would generate unphysical electron transfer between monomers (polarization catastrophe [96]) and the infinite-order induction energy $\tilde{E}_0 - E_0$ would not vanish at large R.

4
Numerical Studies of Convergence Behaviour

4.1
H···H interaction

The regularized approach was first tested on a simple example of two interacting hydrogen atoms [70]. Such a system obviously does not possess any Pauli-forbidden states. However, even for H_2 serious pathologies in the SAPT convergence were observed [45], and these pathologies were successfully eliminated by the regularization technique.

The computations reported in [70] were carried out for the lowest singlet and triplet states of the H···H system at the interatomic distance of 8.0 bohr (corresponding to the minimum of the van der Waals well in the triplet state) and employed the basis set formed by 180 explicitly correlated Gaussian geminals

$$\exp(-\alpha_1|r_1 - R_A|^2 - \alpha_2|r_2 - R_A|^2 - \beta_1|r_1 - R_B|^2 - \beta_2|r_2 - R_B|^2 - \gamma|r_1 - r_2|^2)$$

$$(84)$$

with the nonlinear parameters α_1, α_2, β_1, β_2, and γ optimized variationally for the total energy of the hydrogen molecule. This basis was supplemented by two functions representing the orbital products $1s_A(r_1)1s_B(r_2)$ and $1s_B(r_1)1s_A(r_2)$ with the hydrogenic $1s$ orbital expanded in terms of 60 primitive Gaussian orbitals with even-tempered exponents, so that the hydrogen atom energy in this basis differed from -0.5 by only 2×10^{-14}. The perturbation corrections were computed by expanding the perturbed functions in the basis that diagonalizes the zero-order Hamiltonian H_0 (or H_p in case of the

R-ELHAV method) and using an appropriate spectral representation for the reduced resolvent $\mathcal{R}_0$ ($\mathcal{R}_p$).

The results of [70] confirm that the conventional, non-regularized polarization series converges to the ground-state interaction energy $^1\mathcal{E}$ [46]; however, after approaching quickly the Coulomb energy Q, which differs from $^1\mathcal{E}$ by 31.9841%, the convergence becomes pathologically slow, and the exchange part of the interaction energy is not reproduced to any reasonable extent in a finite-order treatment. These results, including the convergence radius ρ equal to 1.0000000031, are in perfect agreement with those obtained earlier [46] using the explicitly correlated basis of Kołos-Wolniewicz [97].

The conventional SRS series, as for the polarization one, also converges quickly in low orders; however, after reproducing the value of the interaction energy to better than 0.01%, the convergence deteriorates dramatically. This fact is understandable since the RS and SRS expansions possess the same convergence radius. In case of the singlet state, the SRS series converges to the exact interaction energy $^1\mathcal{E}$. For the triplet state this is not possible since the RS expansion for the wave function, from which the SRS energy corrections are calculated (Eq. 12), converges to the fully symmetric singlet function which is annihilated by the antisymmetrizer. As a result, the infinite-order SRS treatment for the lowest triplet state of H_2 yields only the so-called *apparent interaction energy* which differs from the exact value of the triplet energy $^3\mathcal{E}$ by 0.012% [70].

The study of [70] shows also that the regularization of the Coulomb potential removes all the pathologies in the convergence behavior of the RS and SRS theories. The stronger the regularization (the smaller the value of the parameter η), the faster the R-RS expansion approaches its limit. For any finite value of η, the R-RS series converges smoothly (unlike in the case of LiH, as we will see in the next subsection). Comparison of the results obtained with the one-electron regularization and with the full regularization [70] demonstrates that it is the one-electron, attractive part of the perturbation that is responsible for the convergence problems in SAPT. Regularization of the electron-electron repulsion does not change the results significantly. In fact, the calculated values of the convergence radius ρ are the same for both regularization algorithms [70].

Obviously, the limit $\mathcal{E}_p$ of the regularized polarization expansion depends on the value of the regularization parameter η. Fortunately, this dependence is rather weak and for a wide range of η the value of $\mathcal{E}_p$ is very close to the Coulomb energy Q. Similarly, the limit $\mathcal{E}_{R-SRS}$ of the regularized SRS expansion exhibits only weak dependence on the value of η, both for the singlet and triplet state (cf. Table 4 of [70]). Even for quite a strong regularization corresponding to $\eta = 5$, one can recover the exact interaction energy to better than one percent from the R-RS wave function that does not include the effects of V_t. Moreover, the value of $\mathcal{E}_{R-SRS}(\eta)$ can be successfully approximated by a finite sum of the R-SRS energy corrections: the regularized SRS series, unlike the non-regularized one, converges quickly and smoothly.

When one knows the function ψ_P, the small part of the interaction energy that is missing in the infinite-order R-SRS energy $\mathscr{E}_{\text{R-SRS}}$ can be easily recovered by means of the R-ELHAV (or R2-ELHAV) expansion of Sect. 3.3. This expansion converges really fast and, unlike the non-regularized ELHAV series suffering from the wrong asymptotic behavior, gives very accurate results already in a low-order treatment. The rapid high-order convergence of the R-ELHAV expansion results from large values of the convergence radius ρ. In the whole range of η studied in [70], ρ is greater than two, and it increases rapidly when η increases, i.e. when a larger part of the interaction is already included in ψ_P. Switching from the one-electron regularization to the full one does not significantly affect the convergence radii of the R-ELHAV series.

The low-order R-ELHAV and R2-ELHAV energies are similar, although the R-ELHAV results are consistently somewhat more accurate [70]. The success of a low-order R-ELHAV approach is, however, a little paradoxical since the expression for the first-order R-ELHAV energy

$$E^{(1)}_{\text{R-ELHAV}} = N_P \left\langle \psi_P | V_t \mathcal{A} \psi_P \right\rangle \tag{85}$$

involves a one-electron operator only and is completely different from the well-established Heitler-London formula for the leading contribution to the exchange energy [19]. On the other hand, the good accuracy of low-order R2-ELHAV results is well understood since $E^{(1)}_{\text{R2-ELHAV}} = \mathscr{E}_{\text{R-SRS}}$, and higher R2-ELHAV corrections provide an improvement to the already accurate infinite-order R-SRS energy.

4.2
Li···H Interaction

Interacting lithium and hydrogen atoms are the simplest system for which the convergence of the polarization or SRS expansions is destroyed by the Pauli-forbidden continuum in which the physical ground state is submerged. The numerical studies for this system were performed in [57, 69, 71], and [72]. These studies have shown that the regularization of the Coulomb potential leads to expansions that are both convergent and asymptotically correct to any order, not only for the distances around the van der Waals minimum for the triplet state, but also in the region of the chemical minimum for molecular, singlet LiH.

Conventional SAPT Expansions

To make a high-order perturbation treatment computationally feasible, all the numerical calculations of [57, 69], and [72] have been carried out using a rather moderate basis set of 32 Gaussian orbitals. The orbital exponents have been taken from [98] (Li) and [48] (H) and augmented by a set of diffuse functions optimized for the dispersion interaction to obtain a realistic description of the interaction energy in the van der Waals minimum region.

To provide reference values for the interaction energies obtained with SAPT, the full configuration interaction (FCI) calculations for the lowest singlet ($[\lambda] = [22]$) and triplet ($[\lambda] = [211]$) states of LiH, as well as for the unphysical resonance state ($[\lambda] = [31]$) asymptotically degenerate with the former two, were performed [57] for $10 \leq R \leq 20$ bohr. The singlet and resonance potential curves are negative for this range of R (the singlet state exhibits a chemical minimum at $R = 3.015$ bohr) whereas for the triplet state there is a shallow van der Waals minimum at $R = 11.5$ bohr, and the curve passes through zero at about 10.3 bohr. Interestingly enough, it has been found that the position of the resonance state is approximated extremely well by a weighted average of the physical energies, $^{[31]}\mathcal{E} \approx \frac{2}{3}{}^{[22]}\mathcal{E} + \frac{1}{3}{}^{[211]}\mathcal{E}$, cf. the last two columns of Table I in [57]. The theoretical basis of this approximate equality is not clear at the moment.

The results of [57] show that, as predicted by Adams [52–55], the RS, SRS, and HS expansions diverge. In low order, however, the SRS results are quite accurate, although the accuracy of a second-order treatment is somewhat worse than that obtained for the interactions of typical closed-shell systems [20]. In fact, the conventional SRS theory is capable of providing really accurate results only when one goes to somewhat higher orders. As shown in Fig. 4 of [57], the 20th-order SRS treatment for the triplet LiH is in perfect agreement with the FCI values for the whole range of distances considered. In even higher orders, the divergence starts to show up, and the 30th-order SRS results are significantly less accurate for small R. The best way to obtain really accurate results using the SRS approach is to sum the corrections $E_{SRS}^{(n)}$ until they start to grow in absolute value. This is the standard method of obtaining a sum for a series that converges asymptotically.

Reference [57] also reported results obtained with the $1s^2$ core of Li frozen. Freezing the core of the lithium atom has a very little effect on low-order perturbation corrections. In high orders, however, a significant difference is observed between the frozen core results obtained with and without the inclusion of the configuration state functions of $1s^3$ occupancy. When these functions are included, the high-order behavior of the SRS series mimics very well that observed without freezing the core. When the $1s^3$ functions are removed from the basis set, the high-order perturbation corrections change their behavior and the SRS series appears to converge, although extremely slowly. As in the case of two hydrogen atoms (Sect. 4.1), the limit of the SRS series for the triplet state is very close but slightly different from the supermolecular interaction energy. This situation could be expected because the frozen-core RS series for the wave function converges to a state of different permutational symmetry so the SRS energy expression (Eq. 59 in the limit $\eta \to \infty$) takes the form 0/0 at $\zeta = 1$. The results of [57] clearly show that it is the bosonic $1s^3$ state of the lithium atom and the associated Pauli-forbidden continuum of the perturbed Hamiltonian that are responsible for the divergence of the SRS perturbation series for LiH. Removing this continuum makes the SRS expansion convergent; however, it does not improve the rather moderate low-order convergence rate of the perturbation series.

The formal similarity of the SRS and MSMA theories (Sect. 2.2) leads to almost the same convergence behavior of these two methods [69]. As in the case of the SRS expansion, the divergence of the fully correlated theory turns into a moderately fast convergence when the frozen-core approximation with the neglect of $1s^3$ configurations is applied. The close similarity of the SRS and MSMA results is further evidenced by the convergence radii of these series [69]. Thus, the MSMA theory does not provide any improvement over the simple SRS approach. As a matter of fact, the same is true for the much more complicated HS theory [57] – the SRS and HS results are practically the same in low orders, and in higher orders the HS expansion diverges even more rapidly than SRS. Freezing the core of the lithium atom has a similar effect on the SRS and HS corrections; however, when the $1s^3$ configurations are absent in the basis set, the HS series, unlike SRS, converges to the exact supermolecular interaction energy for the triplet state [57]. Divergence of the HS theory, predicted by Adams [55, 68], may be viewed as disappointing since this method was exhibiting the best performance among all investigated methods for systems involving one- and two-electron monomers [47–49]. However, even if the HS theory were convergent, its complex multistate structure would make it extremely difficult to use for larger systems.

The results of [69] show that the conventional, non-regularized ELHAV, JK, and JK-1 expansions converge despite the presence of a Pauli forbidden continuum. However, the wrong asymptotics of the second-order ELHAV energy spoils dramatically its low-order convergence rate, especially for larger intermonomer distances. The JK expansion, exhibiting correct asymptotics in second order, is far better. One may raise a question whether improving further the asymptotic properties while maintaining the strong symmetry forcing, i.e. going from JK to JK-1, results in further improvement of the low-order SAPT convergence rate. The answer is obviously yes for very large intermonomer distances where the potential curve is perfectly described by a truncated C_n/R^n expansion. However, even for R as large as 20 bohr, the difference between the low-order JK and JK-1 energies is not really significant, and a simple SRS theory is almost identical to JK-1 in low orders [69]. For smaller distances, the JK-1 results are not much better than the JK ones. Surprisingly, sometimes they are even slightly worse, as it is the case for the triplet LiH at $R = 11.5$ bohr. This means that there is no point in trying to introduce further improvement to JK by improving the function used to initiate the iterative process defined by Eqs. 10 and 11, as, for example, in the JK-2 approach outlined in [69].

The results of [69] also show that the ELHAV and JK expansions converge uniformly better for the triplet than for the singlet state. This observation is further supported by the convergence radii of ELHAV/JK/JK-1 series given in Table II of this reference. These convergence radii were obtained by extrapolating the d'Alembert ratios or by fitting the large-order energies to the formula [99]

$$E^{(n)} \approx C\rho^{-n} \frac{J_0((n-\frac{3}{2})\theta) - J_0((n+\frac{1}{2})\theta)}{2n-1} , \tag{86}$$

where θ is an argument of the branch points $\rho e^{\pm i\theta}$ ($\theta \ll 1$) determining the convergence radius, J_0 is the standard Bessel function, and C is some constant. It has been found that in the considered range of distances R, the convergence radii for the ELHAV, JK, and JK-1 methods were identical to the number of digits computed. At $R = 11.5$ bohr (i.e. at the bottom of the van der Waals well), ρ is equal to 1.214 for the singlet and 1.600 for the triplet state. In fact, these radii are not exactly identical but the differences between them are visible only at smaller interatomic separations [69]. It may also be noted here that in the case of these three convergent expansions, both algorithms for freezing the core were found to give results very similar to the fully correlated calculations.

The significant difference between the convergence rate of ELHAV/JK/JK-1 approaches for the singlet and triplet state of LiH can be understood by looking at the behavior for $R \to \infty$. One can show, using somewhat heuristic arguments, that at this limit the convergence radius can be expressed as

$$\rho = \frac{N_0}{N_0 - 1} , \tag{87}$$

where $N_0 = \langle \phi_0 | \mathcal{A}\phi_0 \rangle^{-1}$. The limit of N_0 for $R \to \infty$ is determined by the weight of the unit operator in the symmetry projector $\mathcal{A}$. This weight is equal to 1/4 for the singlet state and 3/8 for the triplet state, cf. Eq. (36) of [57], which gives $\rho = 1.333$ for the singlet and 1.600 for the triplet state. It is not clear why the triplet limit is reached already for the smallest distance considered, whereas even for $R = 20$ bohr the singlet convergence radius differs significantly from its asymptotic value. Equation 87 appears to hold also for the H$\cdots$H interaction and the AM theory, as recently found by Adams [100], as well as for the HeH system.[2]

Some information about the convergence of the non-regularized AM and SAM expansions (Sect. 2.3) has been given in [72] for the case of the triplet state at the van der Waals minimum ($R = 11.5$ bohr). The results reported in this reference were obtained with the offset parameter D equal to zero, however, as shown earlier in Table 2, different choices of D lead to very similar results. The low-order convergence of the pure AM theory appears to be the worst of all the SAPT methods considered. Significant improvement is achieved by the symmetrization of the energy expression, and the

---

[2] Przybytek M et al., to be published

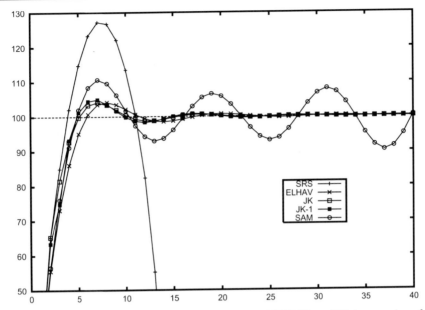

Fig. 2 Convergence of the non-regularized SRS, SAM, ELHAV, JK, and JK-1 expansions for the singlet state of LiH at $R = 3.015$ bohr. The numbers displayed are percentages of the FCI interaction energy recovered in the nth order perturbation treatment

SAM energies are much more accurate, although still not superior (and in fact, almost equal) to those resulting from the ELHAV expansion, let alone the JK one. Analogous results were obtained in the case of the singlet state. The same close similarity between the SAM and ELHAV energies, which could have been expected on the grounds of the formal similarity of these methods (cf. Sect. 2.3), was obtained for the interaction of a triplet helium atom with a hydrogen atom [87]. This similarity disappears when one goes to shorter intermonomer distances, as illustrated in Fig. 2 by the results of non-regularized ELHAV and SAM expansions for the singlet state at the chemical minimum distance of 3.015 bohr. In this region the superiority of ELHAV is clear; in fact, the SAM expansion was found to diverge in this case. The EL-HAV, JK, and JK-1 series remain convergent for such a short intermonomer distance, and in low orders JK and JK-1 give slightly more accurate results than ELHAV.

Regularized SAPT Expansions

The lowest eigenvalue $\mathcal{E}_p$ of the regularized Hamiltonian $H_0 + V_p$ for the hydrogen dimer was found to be very close to the Coulomb energy Q defined as the weighted average of the energies of all asymptotically degenerate states (including the Pauli forbidden ones) with weights proportional to the exchange degeneracy of these states. Comparison of $\mathcal{E}_p(\eta)$ and Q for LiH at a few intermonomer distances is presented in Fig. 3. An analogous comparison of the infinite-order R-SRS energy $\mathcal{E}_{R\text{-}SRS}(\eta)$ and the FCI interaction energy for the triplet state of LiH has been made in Fig. 3 of [72]. These figures show that, unlike for the hydrogen dimer, the regularization must be sufficiently strong to make $\mathcal{E}_p$ resemble Q. There exists a critical value η_c (or ω_c for the SNC regularization) at which the curve $\mathcal{E}_p(\eta)$, and the corresponding eigen-function $\psi_p(\eta)$, undergo a dramatic change of character. Above η_c, the value of $\mathcal{E}_p(\eta)$ exhibits a steep fall towards the energy of the mathematical ground state of H in the space $\mathcal{H}_A \otimes \mathcal{H}_B$, belonging to the Pauli-forbidden [31] sym-metry. The wave function $\psi_p(\eta)$, resembling ϕ_0 for $\eta < \eta_c$, for $\eta > \eta_c$ changes its character and starts to resemble the mathematical ground-state function. This result shows that to have a chance of obtaining convergent regularized SAPT expansions one must regularize the potential strongly enough, i.e. one must set $\eta < \eta_c$. It is interesting to note that the critical value η_c, equal to 5.43, is independent of R for the range of distances considered. The $R \to \infty$ limit of η_c can be calculated from monomer properties [87]. This limit amounts also to 5.43 in the basis set used for the calculations in [72]. Figure 3 also contains the curve $\mathcal{E}_p(\omega)$ obtained using the SNC regularization at $R = 11.5$ bohr. The critical value ω_c, equal to about 1.6, is significantly smaller than η_c. The rea-

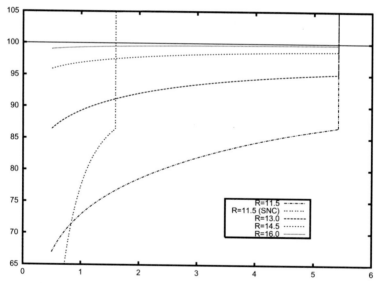

Fig. 3 Dependence of the infinite-order R-RS energy $\mathcal{E}_p$ on the value of the regulariza-tion parameter η (ω) for a few intermonomer distances R. The curve marked "SNC" has been obtained using the SNC regularization, for all the other curves the Gaussian regu-larization was used. The numbers displayed are percentages of the FCI Coulomb energy Q calculated for the same distance R

son for $\omega_c \ll \eta_c$ is that if $\eta = \omega$, the SNC regularization is significantly milder at short distances from the nucleus, cf. Fig. 1 of [70], so the Gaussian regularization leads to a much smaller volume of the regularized Coulomb well. The results of Fig. 3, and of Fig. 3 in [72], show that it is not really the removal of the negative singularities from V that is crucial for the success of the regularized SAPT. It is rather the breaking of the permutational symmetry of the full Hamiltonian $H_0 + V$ and the weakening of the Coulomb attraction between electrons of one monomer and the nucleus of the other one that is really relevant. Therefore, the procedure of [70–72] and [87] could also be referred to as a "short-range attenuation" rather than a "regularization" of the Coulomb potential.

Even when the regularization parameter is smaller than its critical value, the results of Fig. 3 show that the infinite-order energies $\mathcal{E}_p(\eta)$ and $\mathcal{E}_{R\text{-SRS}}(\eta)$ can still be quite different from the reference FCI values. Only for larger R do both $\mathcal{E}_p$ and $\mathcal{E}_{R\text{-SRS}}$ become very accurate. This result might be expected since both these quantities exhibit the same asymptotic behavior as the exact interaction energy. However, as demonstrated in [72], the description of exchange interactions by the infinite-order R-SRS energy also improves when R increases. As a measure of the accuracy of the exchange energy (in the triplet state), one can choose the parameter

$$\Delta = 100\% \times \left| \frac{[211]\mathcal{E}_{R\text{-SRS}} - [211]\mathcal{E}_{FCI}}{[211]\mathcal{E}_{FCI} - [22]\mathcal{E}_{FCI}} \right|, \tag{88}$$

i.e. the percentage error of $\mathcal{E}_{R\text{-SRS}}$ relative to the exponentially vanishing singlet-triplet splitting. The value of Δ is found to decrease with increasing R, however, this decrease is slow and it is not clear from the results obtained thus far (also for the interaction of a triplet helium atom with a hydrogen atom [87]) whether Δ goes to zero for $R \to \infty$. It is worth noting that the SNC regularization, despite having a smaller critical value of the regularization parameter than the Gaussian one, performs similarly to the latter as far as the accuracy of $\mathcal{E}_p$ ($\mathcal{E}_{R\text{-SRS}}$) for $\omega < \omega_c$ is concerned. This fact is demonstrated by the results in Fig. 3 for $R = 11.5$ bohr.

Convergence of the regularized RS and SRS expansions is presented in Tables 3 and 4. As expected, the condition $\eta < \eta_c$ is necessary for the convergence of the R-SRS series. As η increases beyond its critical value, the R-SRS expansion diverges faster and faster, and in the limit $\eta \to \infty$ it becomes identical with the non-regularized SRS theory (having the convergence radius of 0.7167 at $R = 11.5$ bohr [69]). As η decreases below η_c, the SRS series starts to converge smoothly to its limit $\mathcal{E}_{R\text{-SRS}}$, and for a wide range of values of the regularization parameter the R-RS convergence radius is significantly greater than unity. However, the condition $\eta < \eta_c$ does not guarantee the convergence of perturbation series. If η is too small, the R-SRS series starts to diverge in an oscillatory way. This happens when $\eta < 0.7$, or $\omega < 0.16$ in case of the

Table 3 Convergence of the regularized polarization expansion for $R = 11.5$ bohr and both regularizations (Gaussian and SNC). The numbers listed are percent errors of the sum of the first n R-RS corrections with respect to $\mathcal{E}_p(\eta)$ ($\mathcal{E}_p(\omega)$). The values of $\mathcal{E}_p$ (and of the Coulomb energy Q for $\eta = \infty$), given in microhartrees, are displayed in the row marked $\mathcal{E}_p$, whereas the last row lists convergence radii of the regularized polarization series

n	$\eta = 0.5$	$\eta = 1$	$\eta = 5$	$\omega = 0.09$	$\omega = 1$	$\omega = 2.25$	$\eta = \infty$
2	3.0127	−0.3668	−5.4387	−21.1765	−2.5056	−5.2368	−16.6158
3	−2.1651	−1.1629	−3.8041	13.4992	−1.8990	−3.6771	−13.8633
4	0.9617	−0.0591	−1.9792	−9.8130	−0.6486	−1.8921	−10.9174
5	−0.6032	−0.1114	−1.1939	7.5266	−0.3341	−1.1376	−8.9167
6	0.3537	−0.0032	−0.7081	−5.8669	−0.1505	−0.6710	−7.2920
7	−0.2163	−0.0144	−0.4295	4.6026	−0.0745	−0.4050	−5.9860
8	0.1311	0.0000	−0.2623	−3.6206	−0.0361	−0.2459	−4.9206
9	−0.0803	−0.0021	−0.1625	2.8522	−0.0182	−0.1512	−4.0489
10	0.0493	0.0000	−0.1022	−2.2492	−0.0093	−0.0942	−3.3330
15	−0.0064	0.0000	−0.0154	0.7042	−0.0005	−0.0127	−1.2249
20	0.0042		−0.0067	−0.2597	−0.0001	−0.0046	−0.2657
25	−0.0061		−0.0053	0.1672	0.0000	−0.0033	0.5214
30	0.0097		−0.0048	−0.2182		−0.0028	1.9714
40	0.0244		−0.0041	−0.7169		−0.0023	17.1733
50	0.0618		−0.0037	−2.5729		−0.0019	175.8250
60	0.1571		−0.0032	−9.2761		−0.0016	
70	0.4005		−0.0028	−33.4818		−0.0013	
$\mathcal{E}_p$	−34.4311	−39.1431	−44.3471	27.7881	−41.6989	−44.1406	−51.4609
ρ	0.910	1.270	1.013	0.879	1.164	1.019	0.717

SNC regularization. This divergence can be viewed as an artifact of the one-electron regularization. It is probably caused by a mechanism similar to the one that makes the Møller-Plesset perturbation theory divergent for many systems – the so-called backdoor intruder state [101], i.e. the complex function $\mathcal{E}(z)$ has a singularity with a negative real part inside the unit circle. The appearance of such a singularity has been confirmed by recent model studies of Adams [100]; the singularity is likely to show up because, when a strong one-electron regularization is applied, the weak nucleus-electron repulsion present in the operator – V_p cannot compensate for the strong, non-regularized electron-electron attraction from the two-electron part of – V, so the electrons from one monomer fall into Coulomb wells around the other monomer's electrons. This divergence could be avoided if the two-electron part of V were regularized as well. However, the range of values of the regularization parameter η (or ω) for which the R-SRS expansion converges is sufficiently broad for the divergence for small η to bear no practical importance. For the SNC regularization, the convergence patterns are similar to those for the Gaussian one; R-RS diverges for $\omega > \omega_c$ as well as for very small ω. For the values of ω between the divergence regions, the regularized RS and

Table 4 Convergence of the regularized SRS expansion for the lowest singlet and triplet states of LiH, for $R = 11.5$ bohr and both regularizations. The numbers listed are percent errors defined with respect to the sum $\mathcal{E}_{R-SRS}$ of the R-SRS series. The last two rows display values of this sum (in microhartrees) and percent differences δ between $\mathcal{E}_{R-SRS}$ and the FCI interaction energy

n	singlet state				triplet state			
	$\eta = 1$	$\eta = 4$	$\omega = 1$	$\omega = 2.25$	$\eta = 1$	$\eta = 4$	$\omega = 1$	$\omega = 2.25$
2	−2.4559	−6.4502	−4.3457	−7.0680	2.6812	−0.2710	0.9801	−1.0023
3	−0.9980	−3.8285	−2.3196	−4.3374	−1.3357	−2.4001	−1.7090	−2.9252
4	−0.3313	−1.9832	−0.9734	−2.3729	0.2771	−0.6874	−0.1690	−1.0579
5	−0.0936	−1.1253	−0.4729	−1.4091	−0.0682	−0.4175	−0.1350	−0.6598
6	−0.0443	−0.6368	−0.2263	−0.8355	0.0561	−0.1848	−0.0194	−0.3373
7	−0.0122	−0.3657	−0.1119	−0.5008	−0.0003	−0.0873	−0.0022	−0.1791
8	−0.0069	−0.2111	−0.0558	−0.3005	0.0129	−0.0341	0.0080	−0.0867
9	−0.0022	−0.1226	−0.0285	−0.1803	0.0024	−0.0091	0.0084	−0.0368
10	−0.0013	−0.0715	−0.0148	−0.1076	0.0034	0.0027	0.0073	−0.0094
15	0.0000	−0.0041	−0.0007	−0.0028	0.0002	0.0085	0.0018	0.0182
20		0.0011	0.0000	0.0072	0.0000	0.0055	0.0005	0.0163
25		0.0014		0.0079		0.0040	0.0002	0.0143
30		0.0012		0.0075		0.0030	0.0001	0.0128
40		0.0008		0.0064		0.0019	0.0000	0.0105
50		0.0005		0.0054		0.0012		0.0087
60		0.0003		0.0045		0.0007		0.0072
70		0.0002		0.0038		0.0005		0.0059
sum	−68.7700	−73.7824	−71.4893	−74.3672	−15.6509	−15.9859	−15.8418	−16.1331
δ	−18.83	−12.92	−15.62	−12.23	−13.99	−12.15	−12.94	−11.34

SRS series converge, and the accuracy of the low-order corrections, as well as the convergence radii, are similar to the ones obtained with the Gaussian regularization.

The regularized SRS expansion, however quickly convergent, cannot provide a very accurate description of the potential energy curve for distances around the van der Waals minimum. To obtain accurate results one must resort to a theory that takes into account the singular operator V_t. As shown in Table 5, this can be achieved using the R-ELHAV and R2-ELHAV approaches. The R-JK theory gives similar results in this case since the better asymptotic behaviour of its non-regularized version does not matter here as $v_t(r)$ is a short-range potential. Once ψ_p and $\mathcal{E}_p$ are known, the remaining part of the interaction energy is recovered by the regularized ELHAV theory quite effectively, especially for the triplet state. For the singlet state the convergence is somewhat slower; in fact, the regularized theory appears to have the same convergence radius as the non-regularized one. As for $H - H$, the R-ELHAV and R2-ELHAV results are of similar quality.

Table 5 Convergence of the R-ELHAV (columns marked "R") and R2-ELHAV (columns marked "R2") perturbation expansions (in powers of V_t). The numbers displayed are percent errors of the sum of the first n corrections with respect to the FCI interaction energy. Results for $R = 11.5$ bohr and the Gaussian regularization are shown

| | singlet state | | | | triplet state | | | |
| | $\eta = 1$ | | $\eta = 4$ | | $\eta = 1$ | | $\eta = 4$ | |
n	R	R2	R	R2	R	R2	R	R2
1	−17.7309	−18.8325	−8.9794	−12.9165	−13.4716	−13.9868	−12.6418	−12.1456
2	−14.4166	−15.2865	−6.4253	−9.4414	−5.6683	−5.7610	−4.4571	−4.3651
3	−11.7393	−12.4256	−4.6347	−6.8710	−2.4193	−2.4217	−1.6586	−1.6224
4	−9.5667	−10.1100	−3.3516	−4.9893	−1.0424	−1.0302	−0.6245	−0.6080
5	−7.8001	−8.2320	−2.4264	−3.6193	−0.4524	−0.4417	−0.2436	−0.2319
6	−6.3619	−6.7068	−1.7575	−2.6243	−0.1977	−0.1904	−0.0944	−0.0882
7	−5.1900	−5.4666	−1.2733	−1.9024	−0.0869	−0.0825	−0.0381	−0.0341
8	−4.2347	−4.4574	−0.9226	−1.3789	−0.0385	−0.0359	−0.0151	−0.0131
9	−3.4557	−3.6355	−0.6686	−0.9995	−0.0172	−0.0156	−0.0063	−0.0051
10	−2.8203	−2.9657	−0.4845	−0.7244	−0.0077	−0.0068	−0.0026	−0.0020
15	−1.0218	−1.0734	−0.0969	−0.1449	−0.0002	−0.0001	−0.0001	0.0000
20	−0.3703	−0.3889	−0.0194	−0.0290	0.0000	0.0000	0.0000	
25	−0.1342	−0.1409	−0.0039	−0.0058				
30	−0.0486	−0.0511	−0.0008	−0.0012				

In view of the success of the regularized ELHAV method in recovering the contribution to the interaction energy missing in $\mathcal{E}_p$, one may hope that the "all-in-one" R-SRS+ELHAV theory presented in Sect. 3.5 will converge well for the LiH system despite the presence of the Pauli-forbidden continuum. The similarity between the non-regularized ELHAV and SAM theories suggests also that the corresponding R-SRS+SAM approach will perform equally well, at least for larger intermonomer distances where the ordinary AM expansion converges. The results of Table 6, as well as the results of [72], show that both R-SRS+ELHAV and R-SRS+SAM indeed converge rapidly for the van der Waals minimum of the triplet state. In low orders, the R-SRS+ELHAV and R-SRS+SAM energies for the same η are practically identical, in higher orders the R-SRS+ELHAV approach is slightly superior, which is indicated by the estimated convergence radii. For the results given in this table, the smaller the value of η, the faster the series converges. However, as for the regularized SRS theory, if η is too small, the R-SRS+ELHAV and R-SRS+SAM series start to diverge in an oscillatory way. This fact is illustrated by the low-η R-SRS+SAM results displayed in Fig. 4.

The success of the R-SRS+ELHAV theory in reproducing the potential energy curve around the van der Waals minimum of triplet LiH is documented in some detail in [72] where the R-SRS+ELHAV results are compared

Table 6 Convergence of the R-SRS+SAM and R-SRS+SAM+ZI expansions for the triplet state of LiH, for $R = 11.5$ bohr and different values of η. The numbers displayed are percent errors of the sum of the first n corrections with respect to the FCI interaction energy. The row marked "ρ" lists estimated convergence radii of the perturbation series

| n | R-SRS+SAM | | | R-SRS+SAM+ZI | | |
	$\eta = 1$	$\eta = 2$	$\eta = 4$	$\eta = 1$	$\eta = 2$	$\eta = 4$
2	−5.7950	−8.4520	−9.8868	−7.0785	−7.5480	−7.2844
3	−6.5376	−7.9306	−9.3955	−7.8929	−9.1141	−9.5884
4	−2.4007	−4.1470	−5.7491	−3.6992	−4.1591	−4.2570
5	−1.5688	−2.5528	−3.9076	−2.4317	−2.8883	−3.1613
6	−0.6520	−1.4407	−2.5471	−1.3379	−1.5658	−1.7093
7	−0.3755	−0.8270	−1.6662	−0.8018	−0.9724	−1.1220
8	−0.1608	−0.4645	−1.0822	−0.4563	−0.5539	−0.6506
9	−0.0872	−0.2614	−0.7025	−0.2668	−0.3326	−0.4105
10	−0.0383	−0.1464	−0.4559	−0.1538	−0.1941	−0.2474
15	−0.0012	−0.0090	−0.0556	−0.0107	−0.0160	−0.0265
20	−0.0001	−0.0012	−0.0094	−0.0010	−0.0022	−0.0053
25	−0.0001	−0.0005	−0.0032	−0.0002	−0.0007	−0.0022
30	0.0000	−0.0003	−0.0019	−0.0001	−0.0003	−0.0013
40		−0.0001	−0.0011	0.0000	−0.0001	−0.0006
50		0.0000	−0.0007		0.0000	−0.0003
60			−0.0004			−0.0002
70			−0.0003			−0.0001
ρ	1.16	1.12	1.05	1.20	1.13	1.07

with the non-regularized SRS and ELHAV ones for distances ranging from 8 to 16 bohr. The R-SRS+ELHAV method performs similarly to ELHAV for $R = 8$ bohr, and clearly better for larger distances where the non-regularized theory starts to suffer from its incorrect asymptotics. The R-SRS+ELHAV energies are also far more accurate than the SRS ones.

One may ask if the zero-order induction technique of Adams [71] brings about some improvement to the convergence rate of the R-SRS+ELHAV and R-SRS+SAM theories. The results presented in [72] suggest that this is not the case, at least for the triplet state around the van der Waals minimum. The R-SRS+ELHAV+ZI (R-SRS+SAM+ZI) energies are sometimes even slightly worse than the ones of the corresponding R-SRS+ELHAV (R-SRS+SAM) theory, and the convergence radii remain practically unchanged. Thus, the ZI trick does not improve the (already very good) description of the van der Waals well of the triplet LiH.

Except for the non-regularized ELHAV method, all the methods considered here, even the simple SRS theory, provide quite an accurate potential energy curve for the triplet LiH already in the second order (see Fig. 7 in [72]). The most accurate results are obtained from the R-SRS+ELHAV approach; the R-SRS+ELHAV+ZI energies are slightly worse, but both these theories per-

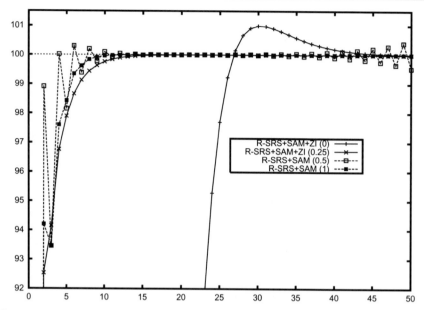

Fig. 4 Percentage of the full CI interaction energy for the triplet state of LiH recovered by the Gaussian-regularized R-SRS+SAM and R-SRS+SAM+ZI expansions through nth order. The interatomic distance is 11.5 bohr, and the numbers displayed in the legend show the corresponding values of η

form far better than SRS, and in the fourth order their accuracy is excellent in the range of distances studied.

There does exist a case when the ZI technique gives a significant improvement – when η is very small. As illustrated by the results in Fig. 4, the R-SRS+SAM+ZI expansions converge for small η and even in the limit $\eta \to 0$, whereas the ordinary R-SRS+SAM series oscillate and are divergent when η is too small. Unfortunately, the accuracy of low-order low-η R-SRS+SAM+ZI energies is really bad, and the advantage of the R-SRS+SAM+ZI expansion over the R-SRS+SAM one (or of R-SRS+ELHAV+ZI over R-SRS+ELHAV) for small η bears no practical importance.

Since the regularization of the Coulomb potential, together with an appropriate symmetry forcing, provided a quickly convergent and fully asymptotically correct description of the van der Waals interaction for triplet LiH, it is interesting to see how the regularized SAPT expansions perform in describing the chemical bond in the singlet LiH. This question was considered by Adams [71], who found that the third-order R-SRS+SAM+ZI approach provides a significant improvement over R-SRS+SAM in this region. Since at small interatomic distances the non-regularized ELHAV expansion is known to exhibit a much better convergence than the SAM expansion (the latter even diverges at the chemical minimum of singlet LiH), one may expect that the R-SRS+ELHAV and R-SRS+ELHAV+ZI approaches will perform better than the corresponding SAM-based methods for this system. The results reported in [72] show that this is indeed the case: the R-SRS+ELHAV expansion remains convergent whereas R-SRS+SAM diverges like its parent SAM method. In fact, the R-SRS+ELHAV approach remains clearly superior to the R-SRS+SAM method for all values of η. The ZI technique of Adams improves the low-order energies of both R-SRS+ELHAV and R-SRS+SAM theories and makes the R-SRS+SAM expansion convergent, at least for some range of η. However, the low-order R-SRS+ELHAV+ZI corrections remain much more accurate than the corresponding R-SRS+SAM+ZI ones [72].

It is very gratifying that the R-SRS+ELHAV approach, both with and without the ZI extension, can successfully describe such different aspects of the interaction phenomenon as the chemical bonding in the singlet state, and the van der Waals attraction in the triplet state of LiH. One may ask, of course, if this conclusion can be transferred to larger systems. We believe that the LiH molecule, unlike H_2 or the ground state of He_2, is a system for which all essential complications plaguing the SAPT treatment of many-electron systems, including the coupling with the Pauli-forbidden continuum, are present. Therefore, we think that the main conclusions reached while investigating the model LiH system will remain valid for larger systems as well, so that the newly developed regularized SAPT may be regarded as a universal theory that is able to accurately describe both chemical and van der Waals interactions – the achievement of a goal set by Eisenschitz and London in the first years of quantum chemistry [22].

5
Extension of the Theory to Many-electron Systems

5.1
Outline of Many-electron SAPT

Sections 2–4 presented the SAPT approach assuming that the Schrödinger equation for the monomers can be solved exactly or nearly exactly. This is not the case for many-electron systems and therefore an extension of SAPT to these systems requires further theoretical developments. In two-body, many-electron SAPT, one uses the following partitioning of the total Hamiltonian:

$$H = F + V + W, \tag{89}$$

where V is the intermolecular interaction operator collecting all Coulomb repulsion and attraction terms between all particles of monomer A and those of monomer B (the same as used in Sects. 2–4), $F = F_A + F_B$ is the sum of the Fock operators for monomers A and B, and $W = W_A + W_B$ is the intramonomer correlation operator with $W_X = H_X - F_X$, X = A or B. The W_X

operators are the same as the Møller-Plesset (MP) fluctuation potentials used in many-body perturbation theories (MBPT) of the electron correlation.

The simplest zero-order wave function in a perturbational approach based on Eq. 89 is the product of monomer Hartree-Fock determinants

$$\Phi_0^{\mathrm{HF}} = \Phi_A^{\mathrm{HF}} \Phi_B^{\mathrm{HF}} . \tag{90}$$

The function Φ_0^{HF} is an eigenfunction of the operator F. The simplest perturbation expansion that one may employ is the RS (polarization) method (extended to the case of two perturbation operators) discussed in Sect. 2. However, as explained in that section, the RS method is not adequate, except for large intermonomer separations. The underlying reason is that the wave functions in this approach do not completely fulfill the Pauli exclusion principle, i.e. the wave functions are not fully antisymmetric with respect to exchanges of electrons [the antisymmetry is satisfied for exchanges within monomers but not between them]. As described in Sect. 2, the antisymmetry requirement can be imposed by acting on the wave functions with the N-electron antisymmetrization operator. This (anti)symmetrization can be performed in many ways and leads to various versions of SAPT. The simplest of them, the SRS method, has been implemented in the many-electron context [24].

The SRS energy corrections are obtained by expanding the expression for the interaction energy in powers of the perturbation operators

$$E_{\mathrm{int}} = \sum_{n=1}^{\infty} \sum_{i=0}^{\infty} E_{\mathrm{SRS}}^{(ni)} = \sum_{n=1}^{\infty} \sum_{i=0}^{\infty} \left(E_{\mathrm{pol}}^{(ni)} + E_{\mathrm{exch}}^{(ni)} \right) , \tag{91}$$

where n denotes the order with respect to V and i denotes the order with respect to W. The last expression splits each $E_{\mathrm{SRS}}^{(ni)}$ component into the polarization part obtained from the RS expansion and the remainder due to the electron exchanges. General expressions for each correction $E_{\mathrm{pol}}^{(ni)}$ and $E_{\mathrm{exch}}^{(ni)}$ were given in [80]. In order to solve the resulting equations for the corrections to the wave function, one usually assumes a finite orbital basis (so-called algebraic approximation), although other approaches are also possible [80]. Derivation of explicit forms of the SRS expressions in terms of two-electron integrals and orbital energies is quite involved. The formulae for several of those components have been given in [23, 25, 102–107].

Each polarization component appearing in Eq. 91 can be related to the physical picture of intermolecular interactions resulting from the long-range expansion of the interaction energy. The first-order corrections $E_{\mathrm{pol}}^{(1i)}$ represent the electrostatic interaction of unperturbed charge distributions and are usually denoted by $E_{\mathrm{elst}}^{(1i)}$. In the second order in V, the contributions are naturally separated into the induction and dispersion components. Also the exchange energies can be related to the corresponding polarization terms. In this way one can, for example, distinguish exchange-dispersion interactions.

The currently implemented level of SAPT represents the four fundamental interaction energy components by the following expansions:

$$E_{elst} = E_{elst}^{(10)} + E_{elst,resp}^{(12)} + E_{elst,resp}^{(13)} \tag{92}$$

$$E_{ind} = E_{ind,resp}^{(20)} + {}^t E_{ind}^{(22)} \tag{93}$$

$$E_{disp} = E_{disp}^{(20)} + E_{disp}^{(21)} + E_{disp}^{(22)} \tag{94}$$

$$E_{exch} = E_{exch}^{(10)} + \epsilon_{exch}^{(1)}(CCSD) + E_{exch-ind,resp}^{(20)} + {}^t E_{exch-ind}^{(22)} + E_{exch-disp}^{(20)} \tag{95}$$

The terms with the subscript "resp" are calculated using monomer orbitals distorted in the field of the interacting partner. The quantity $\epsilon_{exch}^{(1)}(CCSD)$ is the intramonomer correlation contribution to the first-order exchange energy calculated using monomers correlated at the coupled-cluster single and double excitation (CCSD) level, and ${}^t E_{ind}^{(22)}$ and ${}^t E_{exch-ind}^{(22)}$ denote those portions of the second-order intramonomer correlation correction to the induction and exchange-induction energies, respectively, which are not included in $E_{ind,resp}^{(20)}$ and $E_{exch-ind,resp}^{(20)}$. In most applications one adds to the set of SAPT corrections the term:

$$\delta E_{int,resp}^{HF} = E_{int}^{HF} - E_{elst}^{(10)} - E_{exch}^{(10)} - E_{ind,resp}^{(20)} - E_{exch-ind,resp}^{(20)}, \tag{96}$$

where E_{int}^{HF} is the supermolecular Hartree-Fock interaction energy. This procedure involves an approximation since the relation between SAPT and the supermolecular methods can be established rigorously only in the asymptotic limit of large intermolecular separations.

Recently, the complete set of $E^{(30)}$ corrections has been developed [108]. This work allowed us in particular to understand better the role of the term $\delta E_{int,resp}^{HF}$. It has been found that this term should be used only for interactions of polar and polarizable systems, where the induction effects are very large. For systems not fulfilling this criterion, like rare gas atoms or $H_2 - CO$, one should not use $\delta E_{int,resp}^{HF}$.

For large intermonomer separations, the exchange and overlap effects vanish and SAPT interaction energies become equal to those obtained from the multipole expansion [36]. This expansion, often multiplied by some damping factors [109], is applied as the analytic form of functions used to fit the interaction energies computed by SAPT. For small monomers, one can use the expansion in terms of spherical multipole moments, polarizabilities, and dynamic polarizabilities located at the centers of masses of the monomers [36, 110–114]. The coefficients in this expansion are calculated *ab initio* using methods and programs developed by Wormer and Hettema [115, 116]. For larger systems and for the purpose of producing inexpensive potentials for molecular simulations, the center-of-mass expansion is replaced by site-site multipole expansions. The parameters of these expansions can be fitted to the interaction energies of the center-of-mass expansion, but an alternative and a much better approach is to use distributed multipoles [117–121] and polarizabilities [121–126] for the electrostatic and induction energies. Recently, effective methods have been developed for distributing the dynamic polarizabilities, leading to distributed asymptotic dispersion energies [127]. Earlier distributed dispersion approaches [128, 129] are less practical.

5.2
SAPT Based on DFT Description of Monomers

Williams and Chabalowski (WC) [130] have proposed a perturbational approach where the interaction energies are obtained using only the lowest-order, computationally least demanding SAPT expressions, but replacing the Hartree-Fock (HF) self-consistent field (SCF) orbitals and orbital energies by their DFT counterparts, called Kohn-Sham (KS) orbitals and orbital energies. The corrections included were

$$E_{\text{int}} \approx E_{\text{elst}}^{(10)} + E_{\text{exch}}^{(10)} + E_{\text{ind}}^{(20)} + E_{\text{exch-ind}}^{(20)} + E_{\text{disp}}^{(20)} + E_{\text{exch-disp}}^{(20)} . \tag{97}$$

Such an approach is significantly less time consuming than the regular SAPT with high-order treatment of electron correlation. However, the accuracy of the predictions was found to be disappointing [130] even for the electrostatic energy, which is potentially exact in this approach. It has been demonstrated [131] that some deficiencies of the original method stem from an incorrect asymptotic behavior of exchange-correlation potentials. Upon applying an asymptotic correction in monomer DFT calculations, the revised approach was not only able to accurately recover the electrostatic energy, but also the exchange and induction energies. The dispersion energies in the original method of WC were computed from an expression that asymptotically corresponds to the use of uncoupled KS dynamic polarizabilities. Such dispersion energies remained inaccurate even with the asymptotic correction. Misquitta and two of the present authors have proposed [132] a method of computing the dispersion energy that asymptotically corresponds to the use of coupled KS (CKS) polarizabilities. The new method gives extremely accurate dispersion energies, probably more accurate than those predicted by the regular SAPT at the currently programmed level. We will refer to the KS-based SAPT method with the dispersion energy computed in this way as the SAPT(DFT) approach. This approach turned out to be much more accurate than could have been expected. In fact, there are indications that at least for some systems it may be more accurate than SAPT at the currently programmed level [131–134]. A similar approach has been independently developed by Hesselmann and Jansen [135–137].

In the SAPT(DFT) method, the partitioning of Eq. 89 is replaced by

$$H = K + W^{\text{KS}} + V \tag{98}$$

where $K = K_A + K_B$ is the sum of the KS operators (counterparts of F_X), and W^{KS} may be formally defined as $W^{KS} = H_A + H_B - K$. In SAPT, several powers of the W operator, as seen in Eqs. 92–95, have to be included in the expansion to achieve a reasonably accurate description of intermolecular interactions. Since SAPT(DFT) includes only the components listed in Eq. 97 [with the induction and dispersion energies in the CKS versions], SAPT(DFT) effectively neglects the operator W^{KS} and uses the Hamiltonian $H^{KS} = K + V$. Only in the CKS dispersion and induction energies are the effects of W^{KS} partly included (those corresponding to response effects in Hartree-Fock based approaches). The first hint that this approach can work is the fact that if KS orbitals are used in the expression for the electrostatic energy of zero order in W, $E_{elst}^{(10)}$, one gets the exact value of the fully correlated electrostatic energy, $E_{elst}^{(1)}$, provided that DFT gives the exact electron density for a given system. The DFT-based SAPT method in the uncoupled KS version is equally time-consuming as regular SAPT in zero order with respect to W [this level of SAPT will be denoted by SAPT(0)]. The coupled version is somewhat more time-consuming, but still the requirements are much closer to those of SAPT(0) than to those of SAPT at the level of Eqs. 92–95. Thus, the major criterion for evaluation of the effectiveness of SAPT(DFT) is the increase in accuracy of predictions compared with SAPT(0).

Although the original formulation of the SAPT(DFT) method [130] produced interaction energies which were not accurate enough to be competitive with SAPT(0), further modifications of this method not only reversed this relation but made SAPT(DFT) results competitive with the regular SAPT results obtained with high-level treatment of the electron correlation. As mentioned above, the first reason for the initial poor performance was the wrong asymptotic behavior of exchange-correlation potentials in KS equations and this problem can be fixed by applying the asymptotic correction [131]. Then the electrostatic, induction, and exchange energies were reproduced with astounding accuracy and only the (uncoupled KS) dispersion energy was not sufficiently accurate. We will now describe a solution to this problem. The exact (all orders in W) second-order dispersion energy is given by

$$E_{disp}^{(2)} = \sum_{k \neq 0} \sum_{l \neq 0} \frac{|\langle \Phi_0^A \Phi_0^B | V \Phi_k^A \Phi_l^B \rangle|^2}{E_0^A + E_0^B - E_k^A - E_l^B} , \tag{99}$$

where Φ_i^X and E_i^X are the exact eigenfunctions and eigenvalues of H_X. This energy can be represented by an alternative expression [77, 138–140], the so-called generalized Casimir-Polder formula:

$$E_{disp}^{(2)} = -\frac{1}{2\pi} \int_0^{\infty} \int \int \int \int \alpha_A(r_1, r_1'|iu)\alpha_B(r_2, r_2'|iu) \frac{dr_1 dr_2}{|r_1 - r_2|} \frac{dr_1' dr_2'}{|r_1' - r_2'|} du , \tag{100}$$

where

$$\alpha_X(r, r'|\omega) = 2 \sum_{m \neq 0} \frac{E_m^X - E_0^X}{(E_m^X - E_0^X)^2 - \omega^2} \left\langle \Phi_0^X | \hat{\rho}(r) | \Phi_m^X \right\rangle \left\langle \Phi_m^X | \hat{\rho}(r') | \Phi_0^X \right\rangle \qquad (101)$$

is the frequency-dependent density susceptibility (FDDS) of monomer X computed at frequency ω. The symbol $\hat{\rho}(r)$ stands here for the electronic density operator $\hat{\rho}(r) = \sum_i \delta(r - r_i)$, the summation extending over all electrons of the considered molecule. For real ω, $\alpha(r, r'|\omega)$ describes the linear change of electronic density at r under the influence of a one-electron perturbation localized at r' and oscillating with frequency ω. FDDSs are closely related to the dynamic polarizabilities. It should be stressed that the dispersion energy expressions, Eqs. 99 and 100, account not only for the asymptotic dipole-dipole ($1/R^6$) term but also include the effects of all higher instantaneous multipoles as well as the short-range contributions resulting from the overlap of monomer charge distributions [19, 140].

If the wave functions Φ_0^X in Eq. 101 are replaced by appropriate HF/KS determinants, Φ_m^X by singly-excited HF/KS determinants, and the differences $E_m^X - E_0^X$ by the corresponding HF/KS orbital excitation energies, one obtains the uncoupled HF/KS (UCHF/UCKS) FDDSs and the corresponding expression for dispersion energy reduces to the expression for $E_{\text{disp}}^{(20)}$. An obviously better option is to use FDDSs computed using time-dependent DFT (TD-DFT). This choice defines the CKS dispersion energy.

5.3
SAPT/SAPT(DFT) Computer Codes

The computer codes evaluating the SAPT expressions described in Sect. 5.1 have been developed over the last 20 or so years. The current version, named SAPT2002 [24], is available for downloading from the web free of charge (see http://www.physics.udel.edu/~szalewic/SAPT/SAPT.html). The codes are used by about 150 research groups around the world.

The SAPT codes have been parallelized in the period 1997–2004. This work has generated an efficient and portable parallel implementation of a highly correlated electronic structure code suitable for both shared- and distributed-memory parallel architectures. The code features good parallel performance even on Linux clusters with slow communication channels and distributed scratch disk space. Other systems that the parallel SAPT was developed on are IBM SP3/4 and SGI O3K. The tests performed on platforms with common scratch disk space, shared between all processors, show that the competition for I/O bandwidth is the main reason for performance deterioration of I/O-bound codes on such platforms. The parallel implementation of the CCSD method within the SAPT codes was, to our knowledge, the first one efficiently running on more than 32 processors.

The SAPT program scales well up to about 128 processors. The scaling depends on the size of the problem and improves for larger problems. On SP4 and on Linux clusters, the speedups upon doubling the number of processors are generally around 1.8 for the whole range of processors used, both for the integral transformation part and the evaluation of many-body sums. For example, the water dimer calculation in the basis of 430 functions speeds up on SP4 1.7 times upon increasing the number of processors from 32 to 64.

The SAPT(DFT) method is algorithmically almost identical to the SAPT(0) subset of SAPT. In fact, the same code is used for most parts of calculations. The only exceptions are the CKS-type terms whose CHF counterparts in SAPT are computed from different types of formulae. If the uncoupled KS dispersion energies are used, the calculation of the expressions for the corrections listed in Eq. 97 requires a small fraction of the time needed for the calculations of all higher-order corrections listed in Eqs. 92–95. For medium-size (few-atomic) monomers and medium-size (triple-zeta) basis sets, the former calculation is about 2 or 3 orders of magnitude faster than the latter one [130, 133]. If the time spent in SCF/DFT calculations and in the transformation is included, the speedup is reduced to about 1 order of magnitude [133]. Of course, the speedup increases with the size of the system and/or basis set. Thus, the DFT-based approach with the uncoupled KS dispersion energy allows calculations for much larger systems than is possible with the regular SAPT (a further extension of system sizes is possible due to the lesser basis set requirements of SAPT(DFT) [133]). Calculations of the CKS dispersion energies make the timing issues more involved. The expression for this energy scales as the sixth power of the system size, but has a very small prefactor so that this scaling is not consequential for the range of systems of current practical interest. However, somewhat paradoxically, the limiting factor becomes the TD-DFT calculation for monomers. In fact, this calculation wipes out a large part of the speedup quoted above, although for larger systems the better scaling of SAPT(DFT) compared with regular SAPT does result in very large performance gains. Significant work has recently been performed on the optimization of codes and the development of new algorithms for SAPT(DFT) [141]. After only some rudimentary optimizations, the SAPT(DFT) codes became about 1 order of magnitude faster than the initial version and calculations could be performed using modest (workstation-level) computer resources for monomers containing about a dozen atoms in basis sets with a couple of hundred orbitals [142, 143]. Regular-SAPT calculations for such systems are very costly. Using significant computer resources, SAPT(DFT) codes could be used for calculations for dimers containing about 20-atom monomers, which could not be investigated with regular SAPT. A major further speedup has recently been achieved by using the density fitting methods [141]. This approach was particularly important for the TD-DFT and transformation stages of calculations. Similar developments have also been achieved by Hesselmann et al. [144]. The new codes make calculations for systems of the size of benzene dimer in medium size basis sets virtually as inexpensive as the standard supermolecular DFT calculations. Although the SAPT(DFT) method scales worse than DFT with system size N: as N^5 vs. N^3, both methods can be reduced to linear scaling for very large systems.

6
Helium

6.1
Helium as a Thermodynamic Standard

Ab initio computed interaction potentials for the helium dimer turned out to be of significant importance in thermal physics. Thermodynamical measurements in industry and science are utilizing international standards of temperature and pressure and the value of the Boltzmann constant. Improvement of the accuracy of these standards is the major goal of the international metrology community. These quantities are interrelated and can be measured in several ways. If the measurements are performed in helium, it is now possible to utilize *ab initio* values of some quantities, such as the density and dielectric virial coefficients, that previously had to be determined experimentally.

The Boltzmann constant k_B (or equivalently the gas constant, $R = N_A k_B$, where N_A is the Avogadro constant known to 0.17 ppm) was measured in argon by Moldover et al. in 1988 with an accuracy of 1.7 ppm [145]. With improvements in technology, the accuracy can be increased to about 1 ppm within a few years.[3] There appears to be a consensus in the metrology community that after the uncertainty of k_B is reduced to 1 ppm or better, this value should be fixed and further work should concentrate on improving the temperature standard. Once the value of k_B is fixed, e.g. $k_B = 1.38065 \times 10^{-23}$ J/K, one kelvin can be defined as the change of temperature leading to a change of $k_B T$ equal to 1.38065×10^{-23} J. Then, no particular way of measuring the temperature would be favored. However, the required reduction of the uncertainty in the value of k_B is not a simple task. In principle, the Boltzmann constant can be determined with any primary thermometer by measuring the product $k_B T$ at the triple point of water (TPW), which is assumed by the current standard, ITS-90, to be $T = 273.16$ K exactly. In practice, there will always be an uncertainty resulting from the realization of TPW (in the current value of k_B, this uncertainty amounts to 0.9 ppm [145]).

The ITS-90 defines the temperature scale using a set of fixed points and interpolates between these points with platinum thermometers. The uncertain-

---

[3] Moldover MR (2005) Personal communication

ties of the standard compared to thermodynamics measurements approach 70 ppm in some regions. In addition to the limits of accuracy, this standard has the disadvantage of being generally inconsistent with thermodynamic variables. Therefore, creation of a new temperature standard is one of the high-priority goals. The conceptually simplest method can be based directly on the equation of state

$$p = RT\rho \left[1 + B(T)\rho + C(T)\rho^2 + ...\right],\qquad(102)$$

where p is pressure, B and C are the second and third virial coefficients, respectively, and ρ is density. If p and ρ are measured and R, $B(T)$, and $C(T)$ are known, this equation determines T. Since R is known to 1.7 ppm, volume and mass measurements giving ρ can be of similar accuracy, and pressure can be measured to a few ppm at low pressures, T could be determined to a better accuracy than given by ITS-90 provided that the virials are known. Such virials can be computed if the interaction potential between helium atoms is known. An accurate potential for the helium dimer [146, 147] based on SAPT was developed in 1996. This potential has been widely used in thermal physics and in other fields. The current most accurate values of $B(T)$ are those from first-principles calculations and have been computed by Janzen and Aziz [148] using the SAPT96 potential for the helium dimer [146, 147] and by Hurly and Moldover [149] using a modification of SAPT96. By comparing with predictions of other *ab initio* works, which gave the depth of the potential at the minimum about 100 mK different from SAPT96, Hurly and Moldover estimated the uncertainty of theoretical $B(T)$ at 300 K to be 2200 ppm [149]. Recently, more accurate calculations of the He_2 interaction energies have been performed for a few intermolecular separations using the supermolecular approach [150, 151]. These calculations are converged to better than 10 mK at the minimum. If such newer *ab initio* calculations [150–152] are taken into account (showing that the SAPT96 potential was within about 50 mK of the current well depth at the minimum), the uncertainty in $B(T)$ can be reduced to about half of the value assumed by Hurly and Moldover [149]. The current goal is to obtain $B(T)$ accurate to about 100 ppm. This accuracy should be more than sufficient for the future standards as $B(T)\rho$ is about 10^{-3} at 300 K. The contribution of $C(T)$ comes with a still lower weight but will also be required for the next generation of measurements. The three-body potential for helium needed to predict $C(T)$ has been computed some time ago [153], showing that many-body effects are very small for helium.

The simplest approach based on Eq. 102 is in practice replaced by measurements involving the dielectric constant of a gas [154]. This leads to the need for the dielectric virial coefficients, discussed below in the context of the pressure standard.

Whereas with the best conventional piston gauges one can measure pressure to a few ppm at low range, the accuracy of the current standard based on such mechanical pistons can be questioned at high pressures. Moldover [155] proposed that the techniques used in thermometry can be inverted to provide a new standard of pressure. This is possible by combining measurements of the dielectric constant of helium with *ab initio* values of the virials and polarizability of helium. The dielectric constant ϵ can be expressed as [155]:

$$\frac{\epsilon - 1}{\epsilon + 2} = A_\epsilon \rho \left[1 + b(T)\rho + c(T)\rho^2 + ... \right] \tag{103}$$

where A_ϵ is the molar polarizability of gas and b and c are respectively, the second and third dielectric virial coefficients. This equation can be combined with the equation of state to give

$$p = \frac{\epsilon - 1}{\epsilon + 2} \frac{N_A}{A_\epsilon} k_B T \left[1 + B(T)^* \frac{\epsilon - 1}{\epsilon + 2} + C(T)^* \left(\frac{\epsilon - 1}{\epsilon + 2} \right)^2 + ... \right] \tag{104}$$

where $B(T)^*$ and $C(T)^*$ are simple algebraic expressions in terms of the density and dielectric virial coefficients and of A_ϵ. The uncertainties of $k_B T$ near TPW and of N_A are below 1.7 ppm. It is expected that $\epsilon - 1$ can be measured to 5 ppm. Recent theoretical work [156, 157] has determined A_ϵ to within 0.2 ppm. Before this work was performed, the uncertainty of the theoretical value of the molar polarizability amounted to 20 ppm, as estimated by Luther et al. [154]. This uncertainty used to be the major limitation of all methods involving dielectric measurements. Now, the standard with accuracy of about 1 ppm can be created if sufficiently accurate values of $B(T)^*$ and $C(T)^*$ can be predicted by theory.

Accurate *ab initio* calculations for the helium dimer and trimer can be used in several other ways in thermal physics. One can use theoretical predictions to calibrate instruments designed to measure the density and dielectric virial coefficients, viscosity, thermal conductivity, speed of sound, and other properties of gases based on comparisons of theory with experiment for helium.

6.2
Towards 0.01% Accuracy for the Dimer Potential

To determine the second virial coefficients to the required accuracy of about 100 ppm, one needs to know the helium dimer potential to a few mK at the minimum. Consider first the interaction of two helium atoms in the nonrelativistic Born-Oppenheimer (BO) approximation. The best published calculations using the supermolecular method have estimated error bars of 8 mK [150, 151]. More recently, improved calculations of this type reached an accuracy of 5 mK [158]. Also, the SAPT calculations have been repeated with increased accuracy, resulting in an agreement to 5 mK with the supermolecular method. Furthermore, the upper bound from four-electron explicitly correlated calculations is 5 mK above the new supermolecular value [159]. All this evidence from three different theoretical models seems to show convincingly that the BO helium dimer interaction energy at the minimum is now known to within a few mK.

At the accuracy level of a few mK near the minimum, effects beyond the BO approximation have to be considered. The adiabatic effects have been computed by Komasa et al. [160] and for $^4He_2$ amount to – 13.2 mK at the minimum. Recently, these results were found to contain small errors and will have to be recomputed. The next effects to be included are the leading relativistic terms ($\sim \alpha^2$). A preliminary value of the relativistic correction at the minimum is + 15.6 ± 0.4 mK [159].

Once the relativistic correction to the helium dimer potential is known, one should consider the QED corrections. A comparison of the relativistic and QED contributions to the polarizability of the helium atom [156, 157] shows that the latter is only slightly smaller in magnitude than the former. Thus, the QED effects on the helium dimer interaction energies may be of the order of a few mK, i.e. non-negligible. A part of the QED effect, the retardation correction, has been calculated for He_2 by several authors for asymptotic separations. In this region, as first shown by Casimir and Polder, the retardation effects change the $1/R^6$ behavior of the potential into $1/R^7$. This asymptotic contribution has to be included in the He_2 potentials in investigations of some effects sensitive to the tails of the potential. For example, the size of the He_2 molecule changes by about 4% upon the inclusion of the retardation effects [147]. The expressions for the QED corrections to energy of the order α^3 are well known; see for example [161]. Some of the terms in these expressions are analogous to those appearing in the calculations of the relativistic corrections and have already been programmed [159]. The so-called Araki-Sucher term [161] and the Bethe logarithm term are more complicated. A complete calculation of the latter term for He_2 is probably not feasible at this time. However, a recent study for H_2 has shown [162] that the R dependence of the Bethe logarithm is very weak. Therefore, it should be sufficient to use its well-known value for the helium atom. This is similar to the finding for the polarizability of helium that the value of the QED correction computed with no approximations [157] agrees to within 1% with the analogous correction estimated by Pachucki and Sapirstein [163] by neglecting the dependence of the Bethe logarithm on the electric field. When the R dependence of the Bethe logarithm is neglected, one finds [159] that the α^3 QED correction to the dimer well depth amounts to – 1.3 ± 0.1 mK – a value somewhat smaller than suggested by the estimate discussed above.

The second dielectric virial coefficient $b(T)$ is 2 orders of magnitude smaller than the density virial coefficient but the existing absolute uncertainties in the values of these coefficients are expected to be similar. Calculations of $b(T)$ were performed by Moszynski et al. using a fully quantum mechanical approach [164, 165] and the collision induced polarizability from SAPT calculations with a moderately large orbital basis set. However, more recent calculations from [166] and [167] disagree with this work to such an extent that the resulting estimated uncertainties are too large from the experimental point of view. It should be noted, however, that the collision-induced Raman spectra predicted theoretically from the SAPT polarizabilities of [164] and [165] agree very well with the experimental data [168–170].

7
Some Recent Applications of Wave-function-based Methods

7.1
Argon Dimer

Argon gas is widely used in physics and chemistry and is also of importance due to its presence in the atmosphere. In particular, it is used in developing standards for thermal physics. The popular, empirical argon dimer potential of Aziz [171] has generally been assumed to be the best representation of this system. However, this potential was not able to fit all data to within experimental error bars [171, 172]. More importantly, neither this potential nor any other potential for argon was able to predict even qualitatively properties dependent on the highly repulsive region of the potential wall [173]. To resolve these problems, a new *ab initio* potential has recently been developed for the argon dimer [174], extending earlier accurate calculations by Slavíček et al. [175]. This potential was based on calculations using the CCSD method supplemented with the noniterative triple excitations contribution [CCSD(T)] in a sequence of very large basis sets, up to augmented sextuple-zeta quality and containing bond functions, followed by extrapolations to the complete basis set limit. The calculations included intermolecular distances as small as 0.25 Å, where the interaction potential is of the order of 4 keV. The computed points were fitted by an analytic expression. The new potential has the minimum at 3.767 Å with a depth of 99.27 cm^{-1}, very close to the experimental values of 3.761 ± 0.003 Å and 99.2 ± 1.0 cm^{-1} [176]. The potential was used to compute spectra of the argon dimer and the virial coefficients. The latter calculations suggest a possible revision of the established experimental reference results. From the agreement achieved with experimental values and from comparisons of the fit with available piece-wise information on specific regions of the argon-argon interaction, one can assume that the potential of [174] provides the best overall representation of the true argon-argon potential to date, although Aziz's potential [171] is probably slightly more accurate near the minimum. To improve the accuracy of *ab initio* predictions, one has to include electron excitations beyond triples and relativistic effects [174].

For the argon dimer, standard-level SAPT calculations overestimate the van der Waals well depth by about 10%. The reason for this discrepancy has been identified recently [108]: the term $\delta E^{\mathrm{HF}}_{\mathrm{int,resp}}$ (see Sect. 5.1) significantly overestimates the 3rd- and higher-order induction and exchange-induction effects for this system. For example, in the aug-cc-pVQZ basis set supplemented by a set of *spdfg* midbond functions, replacing $\delta E_{int,resp}^{HF}$ by the sum $E_{pol}^{(30)} + E_{exch}^{(30)}$ [108] reduces the well depth predicted by SAPT by 7%, resulting in a potential that agrees well with supermolecular CCSD(T) results.

7.2
He – HCl

He – HCl dimer has been of interest for a long time since its spectra may be used to determine the abundance of HCl in planetary atmospheres or in cold interstellar clouds. Recently, a two-dimensional intermolecular PES for the He – HCl complex has been obtained [177] from *ab initio* calculations utilizing SAPT and an *spdfg* basis set including midbond functions. HCl was kept rigid with a bond length equal to the expectation value $\langle r \rangle$ in the ground vibrational state of isolated HCl. In the region of the minimum, the He – HCl interaction energy was found to be only weakly dependent on the HCl bond length, at least as compared with the case of Ar – HF. This result can be attributed to the smaller dipole moment of HCl relative to HF and the subsequently smaller induction energy component in the case of HCl. The calculated points were fitted using an analytic function with *ab initio* computed asymptotic coefficients. As expected, the complex is loosely bound, with the dispersion energy providing the majority of the attraction. The SAPT PES agrees with the semi-empirical PES of Willey et al. [178] in finding that, atypically for rare gas–hydrogen halide complexes including the lighter halide atoms, the global minimum is on the Cl side (with an intermonomer separation 3.35 Å and a depth of 32.8 cm^{-1}), rather than on the H side, where there is only a local minimum (3.85 Å, 30.8 cm– 1). The ordering of the minima was confirmed by single-point calculations in larger basis sets and complete basis set extrapolations, and also using higher levels of theory. Reference [177] has shown that the opposite findings in the recent calculations of Zhang and Shi [179] were due to the fact that the latter work did not use midbond functions in the basis set. Despite the closeness in depth of the two linear minima, the existence of a relatively high barrier between them invalidates the assumption of isotropy, a feature of some literature potentials. The accuracy of the SAPT PES was tested by performing calculations of rovibrational levels. The transition frequencies obtained were found to be in excellent agreement (to within 0.02 cm^{-1}) with the experimental measurements of Lovejoy and Nesbitt [180]. The SAPT PES predicts a dissociation energy for the complex of 7.74 cm^{-1}, which is probably more accurate than the experimental value of 10.1 ± 1.2 cm^{-1}. The characterizations of three low-lying resonance states through scattering calculations can also be expected to be more accurate than the experimentally derived predictions. The analysis of the ground-state rovibrational wave function shows that the He – HCl configuration is favored over the He–ClH configuration despite the ordering of the minima. This is due to the greater volume of the well in the former case. The potential has been used to calculate the spectra of the HCl dimer in helium [181]. The HCl dimer is analogous to the much investigated HF dimer [182], but much more floppy, so that the tunneling splittings in the former case are about an order of magnitude larger.

The ordering of the minima in He – HCl raises the question of general trends in shapes of potential energy surfaces in the Rg – HX family, where Rg denotes a rare-gas atom and X is a halogen atom. Based on the results for He – HCl, single-point SAPT calculations for some systems, and literature data, the authors of [177] investigated this issue. By analyzing the behavior of individual components of the interaction energy, they were able to rationalize the trends within this family and relate them to monomer properties. The changes of monomer properties are very significant across the family: the polarizabilities increase many times as the monomer size increases, whereas the dipole moments of HX decrease. As an effect, the importance of the induction energy – which strongly favors the Rg – H – X configuration – decreases by a factor of 200 as one moves from Kr – HF to He – HI. Thus, the demarcation line for the ordering of the minima lies between these two complexes. One has to take account of the exchange and dispersion interactions for complexes close to this line in order to predict the correct structure.

7.3
He – N_2O

Reference [183] describes SAPT calculations that were performed to determine a two-dimensional potential for the interaction of the helium atom with the nitrous oxide molecule. This system has been very popular in recent years, as shown by publications of two later potentials from two different groups [184, 185]. The reason for this interest is experimental investigations of N_2O embedded in superfluid helium nanodroplets [186]. To estimate the accuracy of SAPT interaction energies, the authors of [183] performed supermolecular CCSD(T) calculations for selected geometries. The *ab initio* interaction energies were fitted to an analytic function and rovibrational energy levels of He – N_2O were computed on the resulting surface. Extensive comparisons were made with a literature *ab initio* He – CO_2 potential and rovibrational states [240] in order to rationalize the highly counterintuitive observations concerning spectra of N_2O and CO_2 in superfluid helium nanodroplets [186], in particular, the degree of reduction of the rotational constants of the molecules between the gas phase and the nanodroplet. Reference [183] argued that the large reduction of the N_2O rotational constant compared to CO_2 is related to the greater potential depth and the resulting greater probability of attaching helium atoms in the former case. Also, the characteristics of the lowest vibrational levels for the two systems are quite different in a way related to the experimental findings. As a byproduct of this work, accurate multipole moments of N_2O have been computed. The quadrupole, octupole, and hexadecapole moments are significantly different from the experimental values and are probably more accurate than the latter.

Based on simple minimizations of potential energy surfaces, the authors of reference [183] predicted the structures of $He_n - N_2O$ clusters. Independently, and virtually simultaneously, similar structures were seen in experiments [187]. The same structures were later found in quantum Monte Carlo calculations by Paesani and Whaley [188] who used the potential of [183]. Although the latter results are more reliable than the static predictions, the underlying physical mechanisms leading to the creation of given types of clusters are more transparent from the minimization work and analysis of the features of PES.

7.4
He – HCCCN

Another molecule investigated in superfluid helium nanodroplets is cyano-acetylene (HCCCN) [189, 190]. This molecule is of interest as a model of elongated species. Five two-dimensional potential energy surfaces for the interaction of He with HCCCN were obtained from *ab initio* calculations using symmetry-adapted perturbation theory and the supermolecular method at different levels of electron correlation [195]. HCCCN was taken to be a rigid linear molecule with the interatomic distances fixed at the experimental "r_0" geometry extracted from ground-state rotational constants. The complex was found to have the global minimum at a T-shaped configuration and a secondary minimum at the linear configuration with the He atom facing the H atom. Two saddle points were also located. There was good agreement between the positions of the stationary points on each of the five surfaces, though their energies differed by up to 19%. Rovibrational bound state calculations were performed for the $^4He - HCCCN$ and $^3He - HCCCN$ complexes. Spectra (including intensities) and wave functions of $^4He - HCCCN$ obtained from these calculations were presented. No experimental spectra have been published for He – HCCCN, so theory may guide future experiments. The effective rotational constant of HCCCN solvated in a helium droplet was estimated by minimizing the energy of $He_n - HCCCN$ for $n = 2$–12, selecting the $n = 7$ complex as giving the largest magnitude of the interaction energy per He, and shifting the resulting ring of He atoms to the position corresponding to the average geometry of the ground state of the He – HCCCN dimer. This estimate was within 4.8% of the measured value [189].

7.5
$H_2 - CO$

The $H_2 - CO$ system is highly important in astrophysics and has been the subject of investigations from several groups. The 1998 SAPT potential predicted spectra of this dimer [192], in excellent agreement with experimental results of McKellar [193, 194], as well as with later experiments by McKellar [195, 196]. The agreement was so good that there was hope that theoretical results can be used to assign the measured, very dense spectrum of the ortho-H_2 – CO complex. The *ab initio* spectrum was actually not sufficiently accurate, but one could expect that if the potential were tuned to the para-H_2 – CO data, the goal could be reached. This turned out not to be the case, and a new four-dimensional energy surface was developed recently [197]. The *ab initio* calculations have actually been performed on a five-dimensional grid and the 4-D surface has been obtained by averaging over the intramolecular vibration of H_2 (the C – O distance was fixed in [197]). Since the goal was to get as accurate results as possible, the supermolecular CCSD(T) method was used. The correlation part of the interaction energy was obtained by performing extrapolations from a series of basis sets. An analytical fit of the *ab initio* set of points has the global minimum of – 93.049 cm^{-1} for an intermolecular separation of 7.92 bohr in the linear geometry with the C atom pointing towards the H_2 molecule. Thus, the potential is significantly shallower than the SAPT potential of 1998. It has been found that the major reason for this discrepancy is that the SAPT potential included an estimate of the induction and exchange-induction energies of the third and higher orders given by $\delta E_{int,resp}^{HF}$. This approximation turned out to work poorly for H_2 – CO. The new potential has been used to calculate rovibrational energy levels of the para-H_2 – CO complex, which agree very well with those observed by McKellar [195]: their accuracy is better than 0.1 cm^{-1}, whereas the 1998 energies agreed to better than 1 cm^{-1}. The calculated dissociation energy is equal to 19.527 cm^{-1} and is significantly smaller than the value of 22 cm^{-1} estimated from the experiment. It turned out that the use of the experimental dissociation energy was the main reason for the tuning of the 1998 potential not achieving the anticipated accuracy. The predictions of rovibrational energy levels for ortho-H_2 – CO have been performed and will serve to guide assignment of the recorded experimental spectra. Also, the interaction second virial coefficient has been calculated and compared with the experimental data. The agreement between theory and experiment here is unprecedented.

7.6
Methane-water

Methane-water interactions are of significant interest. Not only is this system an important model of hydrophobic interactions, but methane-water clathrates (also commonly called methane hydrates) are one of the largest energy resources on Earth. These clathrates are nonstoichiometric mixtures of the two molecules, with the water molecules forming a cage (clathrate) around a methane molecule. Even conservative estimates predict that methane clathrates contain twice the energetic resources of the conventional fossil fuels in world reserves [198]. A 6-dimensional potential energy surface has been developed for interactions between water and methane [199]. The global minimum of the potential, with a depth of 1.0 kcal/mol, was found in a "hydrogen bond" configuration with water being the proton donor and the bond going approximately through the midpoint of a methane tetrahedral face. However, the interaction was shown to have few of the characteristics of typical hydrogen bonds [200]. The SAPT calculations on a grid of about 1000 points were fitted by an analytic site-site potential using previously developed methods [28, 201]. Most of the sites were placed on the atoms. The asymptotic part of the potential was computed *ab initio* using the Wormer and Hettema set of codes for monomer properties [115, 116]. The classical cross second virial coefficient was calculated and agreed well with some experiments but not with others, allowing an evaluation of the quality of experimental results. The potential is now being used in simulations of methane interactions in aqueous solutions.

7.7
Water Dimer

Interactions between water molecules have been the subject of investigations much more often than any other intermolecular interactions, obviously due to the importance of water in all aspects of human life. An elaborate potential for water using IMPT was developed by Millot and Stone [202] and named ASP. Later an improved version was published [203]. SAPT was used to develop a water dimer potential called SAPT-5s [28–30]. The latter potential and its applications in spectroscopy [29, 204, 205] and in simulations of liquid water [31, 32] have been recently reviewed in [20, 21]. Reference [21] also described an improved potential obtained using SAPT(DFT).

A potential for the water dimer with flexible monomers has recently been developed [206]. The resulting potential energy surface is 12-dimensional, which is at the limits of complexity that one can deal with. About half a million points have been computed (compared to about 2500 in [30] using rigid monomers). Fitting these data required a very substantial effort. The potential will be used in calculations of the spectra of water dimer and of the second virial coefficient for water.

8
Performance of the SAPT(DFT) Method

A recent paper [133] is an extension of earlier work [131] and describes the implementation of the SAPT(DFT) version without the coupled Kohn-Sham dispersion energies. Since this version is based only on Kohn-Sham orbitals and orbital energies, it is called SAPT(KS). In addition to the He_2 and $(H_2O)_2$

systems investigated already in [131], the paper describes calculations for Ne_2 and $(CO_2)_2$ dimers at several geometries. In all cases, calculations were performed using several basis sets and a number of DFT functionals. The role of the asymptotic corrections to the exchange-correlation potentials has been investigated. It was shown that the Fermi-Amaldi correction [207] provides the most reliable results. The role of this correction ranges from dramatic in the case of the electrostatic energy of He_2 (order of magnitude improvements in the intramonomer correlation contribution) to rather small for the carbon dioxide dimer at the level of accuracy possible for this system.

It has been found [133] that SAPT(KS) converges much faster with basis set size than the regular SAPT. This is due to the fact that the rate of convergence in the regular SAPT is determined by the slow convergence of the correlation cusps in products of orbitals. Such terms do not appear in SAPT(KS) except for the dispersion energy. However, even the cusps in the dispersion energy are not a problem for SAPT(KS). This is due to the fact that the dispersion energy converges fast already in the regular SAPT provided that the basis set contains diffuse orbitals (possibly optimized for the dispersion energy) and that the so-called midbond functions are used (orbitals placed mid-way between monomers). Reference [133] found that both the dispersion energy and all other terms of SAPT(KS) are very accurate if this type of basis set is used. This result contrasts with regular SAPT, where bases optimal for dispersion energies lead to slow convergence of other terms. It was possible to achieve practically converged SAPT(KS) results for all systems but the carbon dioxide dimer without any need for extrapolations to the complete basis set (CBS) limit. When such extrapolations were performed for the regular SAPT, it was possible to make comparisons free of any basis set artifacts. Interestingly enough, in all cases the agreement between the two approaches was much better at the CBS limit than in finite bases, where the regular SAPT result had significant basis set incompleteness errors.

SAPT(KS) was found [133] to be relatively independent of the DFT functional used. This is in contrast to supermolecular DFT calculations of interaction energies where this dependence is dramatic. All the modern functionals applied gave reasonably close SAPT(KS) components. However, hybrid potentials were generally performing better than the non-hybrid ones. In particular, the functionals PBE0 [208, 209] and B97-2 [210, 211] gave accurate results.

For He_2, accurate benchmarks exist for all the components of the interaction energy [147]. In some cases, the SAPT(KS) components were closer to the benchmark than the regular SAPT components. This shows that the truncation of the expansion in W in SAPT introduces larger errors than the inaccuracies of the current best DFT functionals. For systems larger than He_2, however, the only benchmarks available are those from SAPT. Thus, it is very difficult to answer the question of whether SAPT(KS) components may be more accurate for such systems than the regular SAPT ones. However, in a few cases this was possible. In particular, since recently the electrostatic energy has been available in SAPT at the CCSD level,[4] see also [212]. $E_{elst}^{(1)}$(CCSD) inlcudes a much higher level of electron correlation than does $E_{elst,resp}^{(12)} + E_{elst,resp}^{(13)}$. Some additional estimates could also be performed based on asymptotic comparisons [133]. In all cases, the SAPT(KS) results were closer to the values computed at the higher level than to the regular SAPT results. This shows that – for the electrostatic, first-order exchange, second-order induction and exchange-induction energies – SAPT(KS) is not only approaching but occasionally surpassing the accuracy of regular SAPT at the currently programmed theory level.

Reference [133] provided theoretical justifications for high accuracy of SAPT(KS) predictions for the electrostatic, first-order exchange, and second-order induction energies. For the electrostatic energy, the argument is very simple and has already been given here in Sect. 5.2. The CKS induction energies developed in [133] are analogous to the CKS dispersion energies. As for the dispersion energy [cf. Eqs. 99 and 100], the exact second-order induction energy can be written in terms of the FDDSs, now computed at zero frequency, and of the electrostatic potentials of the unperturbed monomers. In the CKS induction energy, one uses CKS FDDSs and regular DFT electrostatic potentials. Since both quantities are potentially exact in DFT, i.e. would be exact if the exact exchange-correlation potential were known, the induction energy is potentially exact. Since modern density functionals are, for these purposes, reasonable approximations to the exact functional, the CKS induction energies are quite accurate. In fact, somewhat surprisingly, the uncoupled KS induction energies are also very accurate. Reference [133] rationalized this behavior by conjecturing that there is a systematic cancellation between uncoupled and coupled polarizability differences and the respective differences in the overlap effects. Indeed, asymptotically – where overlap effects are small – the CKS induction energies are more accurate.

For the exchange energies, the justification of the good performance of SAPT(KS) is more difficult. An asymptotic expression has been developed [133] for the interaction density matrices which determine the first-order exchange energy in the case of the KS determinants and the exact wave functions. It was shown that in the limit of exact DFT both densities decay in the same way. Thus, at least asymptotically, the first-order exchange energies are potentially exact.

A related manuscript has been devoted to the coupled Kohn-Sham dispersion energies in SAPT(DFT) [134]. The method utilizes a generalized Casimir-Polder formula and frequency-dependent density susceptibilities of monomers obtained from time-dependent DFT. Numerical calculations were performed for the same systems as in [133]. It has been shown that for a wide range of intermonomer separations, including the van der Waals and the short-range repulsion regions, the method provides dispersion energies with

---

[4] Wheatley RJ (2003) unpublished results

accuracies comparable with those that can be achieved using the current most sophisticated wave function methods. The dependence of the CKS dispersion energy on basis sets and on variants of the DFT method has been investigated and the relations were found to be very similar to those for the SAPT(KS) theory discussed above. For the carbon dioxide dimer, the dispersion energy predicted by SAPT(DFT) turned out to be significantly different from that given by SAPT at the level of Eq. 94. An asymptotic analysis strongly suggested that it is the CKS dispersion energy which is more accurate. A further confirmation comes from the fact that whereas SAPT predictions for the CO_2 dimer are significantly different from supermolecular CCSD(T) interaction energies, SAPT(DFT) is in very good agreement with the latter. This finding resolves the long-standing issue [201] of the somewhat unsatisfactory performance of SAPT for this particular system.

If the CKS dispersion energy is combined with the electrostatic and exchange interaction energies from the SAPT(KS) method and the CKS induction energies, very accurate total interaction potentials are obtained. For the helium dimer, the only system with nearly exact benchmark values, SAPT(DFT) reproduces the interaction energy to within about 2% at the minimum and to a similar accuracy for all other distances ranging from the strongly repulsive to the asymptotic region. An accurate SAPT(DFT) potential for this system has also been published by Hesselmann and Jansen [213]. For the remaining systems investigated in [134], the quality of the interaction energies produced by SAPT(DFT) is so high that these energies may actually be more accurate than the best available results obtained with wave function techniques. At the same time, SAPT(DFT) is much more efficient computationally than any method previously used for computing the dispersion and other interaction energy components at this level of accuracy, as discussed in Sect. 5.3.

9
Applications of SAPT(DFT) to Molecular Crystals

One of the most interesting applications of SAPT(DFT) is to predicting structures of molecular crystals. Until recently, these crystals could only be investigated using empirical potentials since, due to the size of typical monomers, SAPT calculations were not practical for such systems.

Molecular crystals are an important class of matter. The current progress in molecular biology and medicine would not be possible if structures of biomolecules were not known from an X-ray analysis of crystalline forms of these compounds. Typical medical drugs are molecular crystals and so are most energetic materials. The ability to predict polymorphic forms of medical drugs would be extremely important in pharmacology. Significant experimental efforts are directed towards creating new materials in the form of molecular crystals with some desired properties.

Many properties of molecular crystals can be predicted theoretically [214–216]. The first step in theoretical investigations of such crystals has to be a determination of interactions (force fields) between constituent molecules. The force field for a crystal can be built from pair and pair-nonadditive contributions computed for isolated dimers, trimers, etc. Such force fields can be obtained empirically or from first-principles calculations. The next step is to determine the low-energy crystal structures applying molecular packing programs. So far, this has been done mostly using empirical force fields. However, the inherent inaccuracies of such force fields limit the predictability of this approach [214–217]. A solution could be to use force fields obtained from first-principles quantum mechanical calculations. Unfortunately, wave-function-based methods are too time consuming for this, whereas, as discussed earlier, the supermolecular DFT approach cannot predict dispersion energies which can be significant for these systems. For a recent demonstration of this fact, see the work of Byrd et al. [218]. Clearly, the solution to this problem is to use SAPT(DFT).

It is hoped that the force fields computed using first-principles methods should be able to predict properties of molecular crystals significantly better than it is currently possible using empirical potentials. In particular, theory could play a very important role in screening *notional* materials, i.e. molecules that may not have been synthesized, but based on their expected structures appear to have desired properties. For such molecules, specific empirical potentials are simply unknown. Generic potentials can be used, but these are not very reliable in predicting crystal structures.

In the subsections that follow, we will describe some pilot SAPT(DFT) calculations for molecules that can be considered models for those forming molecular crystals.

9.1
Benzene Dimer

Solid benzene is one of the most thoroughly investigated molecular crystals as it represents the model system for aromatic compounds. However, there exist no *ab initio* potentials for the benzene dimer. *Ab initio* calculations have been performed only for selected geometries and it is difficult to estimate their accuracy. Recently, SAPT(DFT) was applied to this system [143]. An augmented double-zeta basis supplemented by a set of bond functions was used and calculations were performed for a range of intermolecular separations in the "sandwich" configuration. As it is well known, supermolecular DFT methods fail completely for the benzene dimer, in most cases predicting the wrong sign of the interaction energy [219]. SAPT(DFT) curve agreed very well with the best previous calculations performed by Tsuzuki et al. at the CCSD(T) level and including some extrapolations [220]. Near the minimum, at $R = 3.8$ Å, the two methods predict interaction energies of -1.62 and -1.67 kcal/mol, respectively. However, the residual error of either result (due to basis set truncations and other factors) is somewhat larger than their difference, as indicated by extensive R12-MP2 plus CCSD(T) single-point calculations by Sinnokrot et al. [221] who obtained − 1.81 kcal/mol at 3.7 Å. The latter result agrees very well with preliminary SAPT(DFT) calculations in very large basis sets. Reference [143] also included calculations for Ar_2 and Kr_2, in both cases obtaining excellent agreement with the best existing potentials. The accuracy for all three systems was significantly higher than that of some recent DFT approaches created specifically for calculations of intermolecular interactions [222, 223].

9.2
Dimethylnitramine Dimer

Dimethylnitramine (DMNA) is an important model compound for energetic materials and was investigated by SAPT in the past [224]. Recently, interaction energies were computed for the DMNA dimer containing 24 atoms [142]. In Table 7, the total interaction energies at a near minimum geometry computed using SAPT, SAPT(DFT), and several supermolecular methods are shown. The basis set used was of double-zeta quality with bond functions in monomer-centered "plus" basis set (MC^+BS) form [37] [dimer-centered "plus" basis set (DC^+BS) form in the (counterpoise corrected) supermolecular calculations]. The "plus" denotes here the use of bond functions (in the MC^+BS case, also the use of the isotropic part of the basis of the interacting partner). The regular SAPT results given in Table 7 employ the complete standard set of corrections, in contrast to [224] which used $E_{int}^{HF} + E_{disp}^{(20)}$.

Table 7 Interaction energies (in kcal/mol) for the DMNA dimer, from [142]. The geometry was the near-minimum one denoted by M1 in Table 3 of [224] and the basis set was also taken from that reference. MPn denotes many-body perturbation theory with the MP Hamiltonian

Hartree-Fock	2.25
Frozen-core	
MP2	−7.90
MP4	−7.85
CCSD	−5.31
CCSD(T)	−6.85
All electrons	
CCSD(T)	−6.86
$E_{int}^{HF} + E_{disp}^{(20)}$	−10.58
SAPT (Eqs. 92–95)	−7.36
SAPT(DFT)/PBE0	−6.22
SAPT(DFT)/B97-2	−6.56

Table 8 Individual components of the DMNA dimer interaction energy for SAPT and SAPT(DFT) with PBE0 and B97-2 functionals. Energies are in kcal/mol. The value in parentheses is $E_{disp}^{(20)}$. All data are from [142]

Component	SAPT	PBE0	B97-2
Electrostatic	−10.51	−10.25	−10.05
1st order exchange	18.28	17.43	16.85
Induction	−6.07	−6.54	−6.35
Exchange-induction	4.49	5.02	4.82
Dispersion	−13.50 (−12.83)	−11.83	−11.75
Exchange-dispersion	1.34	1.34	1.30
δE_{int}^{HF}	−1.38	−1.38	−1.38
Total	−7.36	−6.22	−6.56

Table 7 shows first that the higher-order terms neglected in [224] are important for the DMNA dimer and decrease the magnitude of the interaction energy by more than 3 kcal/mol. The value of $E_{int}^{HF} + E_{disp}^{(20)}$ in Table 7 differs from the minimum energy of − 11.06 kcal/mol given in Table 3 of [224] due to the use of the DC$^+$BS vs. MC$^+$BS scheme. SAPT(DFT) gives interaction energies within about 1 kcal/mol of the regular SAPT and about 0.5 kcal/mol of the CCSD(T) method. This constitutes excellent agreement taking into account that both the regular SAPT and CCSD(T) methods are much more computer resource intensive than SAPT(DFT). The SAPT(DFT) calculations were performed with two very different functionals: PBE0 and B97-2, which gave results within 0.3 kcal/mol of each other, showing again that SAPT(DFT) is only weakly dependent on the choice of the functional.

The framework of SAPT provides insights into the physical structure of the interaction energy. Table 8 shows the individual contributions. It can be seen that, as already pointed out in [224], SAPT results do not support the conventional description of interactions of large molecules, which considers only the electrostatic component. Clearly, the first-order exchange and the dispersion energies are actually larger in magnitude than the electrostatic interactions. An attempt to describe the DMNA dimer at the Hartree-Fock level, as it is often done for large molecules, would lead to completely wrong conclusions as the interaction energy at this level is positive. Note also the good agreement between the individual SAPT and SAPT(DFT) components.

10
Transferable Potentials for Biomolecules

Weak interactions between biomolecules govern a significant part of life's processes. With the greatly expanded range of systems that can be investigated *ab initio* as a result of the development of SAPT(DFT), some of these processes become important subjects for computational research. The smallest biomolecules, such as DNA bases, polypeptides, and sugars, are within the range of systems for which the complete potential energy surfaces can be obtained. So far, *ab initio* calculations for such systems have been restricted to single-point calculations [225, 226]. Due to the size of these systems, often even single-point calculations could only be performed at low levels of theory and using very small basis sets. If PESs are available, properties of biomolecules in aqueous solutions can be studied by molecular simulations. Such studies with the use of empirical potentials have been very popular [227]. For some systems experimental data exist also for isolated dimers [228] so that direct comparisons with *ab initio* calculations are possible.

Ab initio methods can also be used in several ways to investigate much larger molecules than one can afford by direct calculations. One avenue is to perform calculations on a fragment of a large biological molecule interacting with another fragment or with a smaller molecule, for example with water. Another way to extend the range of applicability is to use *ab initio* information to construct universal force fields for biomolecules. Such force fields are similar to the currently used empirical ones [229–231] in the sense that the interaction between two arbitrary molecules depends only on predetermined interactions between pairs of atoms belonging to different molecules. A given atom may come in a few "varieties", depending on its chemical surrounding. In the existing biomolecular force fields, this information comes from atomic properties such as polarizabilities, van der Waals radii, and partial charges. In many force fields, some of the parameters, in particular the partial charges, have been obtained from *ab initio* calculations for monomers. In fact, very intensive investigations have been performed to represent electrostatic interactions using the distributed multipole analysis [117, 119, 121]. In recent years, a subset of parameters in force fields has often been tuned by adjusting them within molecular dynamic simulations of some model systems. Such force fields are usually specific for a class of systems, for example proteins.

An ongoing research effort is aimed at proposing an universal force field based on SAPT calculations for a set of model complexes.[5] This project had started before SAPT(DFT) was fully developed, and therefore mostly the regular SAPT approach was used. The future use of SAPT(DFT) will make it possible to increase the size of the model complexes. One component of the new force field, the electrostatic energy, is not parametrized but actually computed as the Coulomb interaction of the charge distributions of the interacting monomers [232]. In this way, the overlap (penetration) effects are fully taken into account and there are no problems appearing related to

---

[5] Volkov A, Coppens P, Macchi P, Szalewics K unpublished results

the divergence of multipole expansions. The monomer (molecular) charge distributions are represented as linear combinations of the so-called pseudoatom densities [233, 234]. The pseudoatom densities have been extracted from *ab initio* molecular densities of a large number of small molecules using a least-squares projection technique in Fourier-transform space [235], and are available in the form of a "databank", with the first applications reported in [236]. For pseudoatoms that are far apart (separations larger than 4–5 Å), the electrostatic energy can be computed from distributed multipoles evaluated directly from pseudoatom parameters [234]. In effect, the calculation of the electrostatic energy is reasonably fast although, of course, not as fast as a summation of point-charge interactions. The remaining terms of the interaction energy are obtained by simultaneous fits of the computed SAPT components for all model systems. Thus, the interaction energy is approximated as[5]

$$E_{int} = E_{elst}^{(1)} + \sum_{a \in A, b \in B} \left[B_a B_b e^{-(C_a + C_b) r_{ab}} - \sum_{n=6,8,10} \frac{A_a^n A_b^n}{r_{ab}^n} \right]. \tag{105}$$

This formula is somewhat reminiscent of the one proposed by Spackman [237–239], but the method of determination of the parameters is completely different. The parameters appearing in the exponential term were fitted to reproduce the exchange energy of Eq. 95 whereas those in front of inverse powers of distances were fitted to the sum of the induction and dispersion energies of Eqs. 93 and 94 for the model set. In practice, some more time-consuming terms in these equations were neglected. A variant of fitting the induction and dispersion energies separately was also explored. There were more than one hundred dimer configurations in the model set. The monomers included amino acids such as: L-serine, L-glutamine, α-glycine, and several other; amino-acid-like compounds such as, for example, DL-norleucine; L-(+)-lactic acid, and benzene. The force field was tested by predicting interaction energies for configurations not included in the model set and for other dimers for which *ab initio* calculations were possible. It was also tested by finding the lattice binding energies for the crystals built of model monomers.

Acknowledgements This research was supported by the NSF grant CHE-0239611 and by an ARO DEPSCoR grant. B.J. acknowledges a generous support from the Foundation for Polish Science.

References

1. Hutson JM (1990) Annu Rev Phys Chem 41:123
2. Cohen RC, Saykally RJ (1991) Annu Rev Phys Chem 42:369
3. Buck U (1975) Adv Chem Phys 30:313

4. Melnick GJ, Stauffer JR, Ashby MLN, Bergin EA, Chin G, Erickson NR, Goldsmith PF, Harwit M, Howe JE, Kleiner SC, Koch DG, Neufeld DA, Patten BM, Plume R, Schieder R, Snell RL, Tolls V, Wang Z, Winewisser G, Zhang YF (2000) Astrophys J 539:L77

5. Allen MP, Tildesley DJ (1987) Computer Simulation of Liquids. Clarendon Press, Oxford

6. Toennies JP, Vilesov AF (1998) Annu Rev Phys Chem 49:1

7. Callegari C, Lehmann KK, Schmied R, Scoles G (2001) J Chem Phys 115:10090

8. Burger A (1983) A Guide to the Chemical Basis of Drug Design. Wiley, New York

9. Naray-Szabo G (ed) (1986) Theoretical Chemistry of Biological Systems, vol 41 of Studies in Theoretical and Physical Chemistry. Elsevier, Amsterdam

10. Autumn K, Liang YA, Hsieh ST, Zesch W, Chang WP, Kenny TW, Fearing R, Full RJ (2000) Nature 405:681

11. Chałasiński G, Szczęśniak MM (1994) Chem Rev 94:1723

12. Chałasiński G, Szczęśniak MM (2000) Chem Rev 100:4227

13. Bartlett RJ (1989) J Phys Chem 93:1697

14. Peréz-Jordá JM, Becke AD (1995) Chem Phys Lett 233:134

15. Chałasiński G, Szczęśniak MM (1988) Mol Phys 63:205

16. Rode M, Sadlej J, Moszyński R, Wormer PES, van der Avoird A (1999) Chem Phys Lett 314:326

17. Sadlej AJ (1991) J Chem Phys 95:6705

18. van Duijneveldt FB, van Duijneveldt-van de Rijdt JGCM, van Lenthe JH (1994) Chem Rev 94:1873

19. Jeziorski B, Moszyński R, Szalewicz K (1994) Chem Rev 94:1887

20. Jeziorski B, Szalewicz K (2003) In: Wilson S (ed) Handbook of Molecular Physics and Quantum Chemistry. Wiley, vol 3, Part 2, Chap. 9, p 232

21. Szalewicz K, Bukowski R, Jeziorski B (2005) In: Dykstra CE, Frenking G, Kim KS, Scuseria GE (eds) Theory and Applications of Computational Chemistry: The First 40 Years. A Volume of Technical and Historical Perspectives, Chap 33. Elsevier, Amsterdam, p 919

22. Eisenschitz R, London F (1930) Z Phys 60:491

23. Rybak S, Jeziorski B, Szalewicz K (1991) J Chem Phys 95:6579

24. SAPT2002: An Ab Initio Program for Many-Body Symmetry-Adapted Perturbation Theory Calculations of Intermolecular Interaction Energies Bukowski R, Cencek W, Jankowski P, Jeziorska M, Jeziorski B, Kucharski SA, Lotrich VF, Misquitta AJ, Moszyński R, Patkowski K, Rybak S, Szalewicz K, Williams HL, Wormer PES, University of Delaware and University of Warsaw (http://www.physics.udel.edu/~szalewic/ SAPT/SAPT.html)

25. Szalewicz K, Jeziorski B (1997) In: Scheiner S (ed) Molecular Interactions – from van der Waals to strongly bound complexes. Wiley, New York, p 3

26. Jeziorski B, Szalewicz K (1998) In: von Ragué Schleyer P et al. (eds) Encyclopedia of Computational Chemistry, vol 2. Wiley, Chichester, UK, p 1376

27. Moszyński R, Wormer PES, van der Avoird A (2000) In: Bunker PR, Jensen P (eds) Computational Molecular Spectroscopy. Wiley, New York, p 69

28. Mas EM, Szalewicz K, Bukowski R, Jeziorski B (1997) J Chem Phys 107:4207

29. Groenenboom GC, Mas EM, Bukowski R, Szalewicz K, Wormer PES, van der Avoird A (2000) Phys Rev Lett 84:4072

30. Mas EM, Bukowski R, Szalewicz K, Groenenboom G, Wormer PES, van der Avoird A (2000) J Chem Phys 113:6687

31. Mas EM, Bukowski R, Szalewicz K (2003) J Chem Phys 118:4386

32. Mas EM, Bukowski R, Szalewicz K (2003) J Chem Phys 118:4404
33. Keutsch FN, Goldman N, Harker HA, Leforestier C, Saykally RJ (2003) Mol Phys 101:3477
34. Patkowski K, Korona T, Moszyński R, Jeziorski B, Szalewicz K (2002) J Mol Struct (Theochem) 591:231
35. Brudermann J, Steinbach C, Buck U, Patkowski K, Moszyński R (2002) J Chem Phys 117:11166
36. Stone AJ (1996) The Theory of Intermolecular Forces. Clarendon Press, Oxford
37. Williams HL, Mas EM, Szalewicz K, Jeziorski B (1995) J Chem Phys 103:7374
38. Chipman DM, Hirschfelder JO (1973) J Chem Phys 59:2838
39. Chałasiński G, Jeziorski B, Szalewicz K (1977) Int J Quantum Chem 11:247
40. Jeziorski B, Szalewicz K, Chałasiński G (1978) Int J Quantum Chem 14:271
41. Chałasiński G, Szalewicz K (1980) Int J Quantum Chem 18:1071
42. Jeziorski B, Schwalm WA, Szalewicz K (1980) J Chem Phys 73:6215
43. Certain PR, Hirschfelder JO, Kołos W, Wolniewicz L (1968) J Chem Phys 49:24
44. Bowman JD (1973) PhD Thesis, University of Wisconsin
45. Ćwiok T, Jeziorski B, Kołos W, Moszyński R, Szalewicz K (1992) J Chem Phys 97:7555
46. Ćwiok T, Jeziorski B, Kołos W, Moszyński R, Rychlewski J, Szalewicz K (1992) Chem Phys Lett 195:67
47. Ćwiok T, Jeziorski B, Kołos W, Moszyński R, Szalewicz K (1994) J Mol Struct (Theochem) 307:135
48. Korona T, Jeziorski B, Moszyński R, Diercksen GHF (1999) Theor Chem Acc 101:282
49. Korona T, Moszyński R, Jeziorski B (1997) Adv Quantum Chem 28:171
50. Hirschfelder JO, Silbey R (1966) J Chem Phys 45:2188
51. Chipman DM, Bowman JD, Hirschfelder JO (1973) J Chem Phys 59:2830
52. Adams WH (1990) Int J Quantum Chem S24:531
53. Adams WH (1991) Int J Quantum Chem S25:165
54. Adams WH (1994) Chem Phys Lett 229:472
55. Adams WH (1996) Int J Quantum Chem 60:273
56. Morgan JD III, Simon B (1980) Int J Quantum Chem 17:1143
57. Patkowski K, Korona T, Jeziorski B (2001) J Chem Phys 115:1137
58. Kutzelnigg W (1980) J Chem Phys 73:343
59. Hirschfelder JO (1967) Chem Phys Lett 1:343
60. van der Avoird A (1967) J Chem Phys 47:3649
61. Peierls R (1973) Proc R Soc London, Ser A, 333:157
62. Amos AT, Musher JI (1969) Chem Phys Lett 3:721
63. Polymeropoulos EE, Adams WH (1978) Phys Rev A 17:18
64. Kutzelnigg W (1978) Int J Quantum Chem 14:101
65. Ahlrichs R (1976) Theor Chim Acta 41:7
66. Jeziorski B, Kołos W (1982) In: Ratajczak H, Orville-Thomas W (eds) Molecular Interactions, vol 3. Wiley, New York, p 1
67. Jeziorski B, Kołos W (1977) Int J Quantum Chem Suppl 1 12:91
68. Adams WH (1999) Int J Quantum Chem 72:393
69. Patkowski K, Jeziorski B, Korona T, Szalewicz K (2002) J Chem Phys 117:5124
70. Patkowski K, Jeziorski B, Szalewicz K (2001) J Mol Struct (Theochem) 547:293
71. Adams WH (2002) Theor Chem Acc 108:225
72. Patkowski K, Jeziorski B, Szalewicz K (2004) J Chem Phys 120:6849
73. Stone AJ, Hayes IC (1982) Faraday Discuss 73:19
74. Hayes IC, Stone AJ (1984) Mol Phys 53:69
75. Hayes IC, Stone AJ (1984) Mol Phys 53:83

76. Hayes IC, Hurst GJB, Stone AJ (1984) Mol Phys 53:107
77. Longuet-Higgins HC (1965) Disc Farad Soc 40:7
78. Hirschfelder JO (1967) Chem Phys Lett 1:325
79. Bloch C (1958) Nucl Phys 5:329
80. Szalewicz K, Jeziorski B (1979) Mol Phys 38:191
81. Claverie P (1971) Int J Quantum Chem 5:273
82. Adams WH (2002) Int J Quantum Chem 90:54
83. Matsen FA (1964) Adv Quantum Chem 1:59
84. Kaplan IG (1975) Symmetry of Many-Electron Systems. Academic Press, New York
85. Murrell JN, Shaw G (1967) J Chem Phys 46:1768
86. Musher JI, Amos AT (1967) Phys Rev 164:31
87. Przybytek M, Patkowski K, Jeziorski B (2004) Collect Czech Chem Commun 69:141
88. Adams WH (2002) J Mol Struct (Theochem) 591:59
89. Korona T, Moszyński R, Jeziorski B (1996) J Chem Phys 105:8178
90. Herring C (1962) Rev Mod Phys 34:631
91. Ewald PP (1921) Annalen der Physik, Ser. 4 64:253
92. Panas I (1995) Chem Phys Lett 245:171
93. Dombroski JP, Taylor SW, Gill PMW (1996) J Phys Chem 100:6272
94. Sirbu I, King HF (2002) J Chem Phys 117:6411,
95. Patkowski K (2003) PhD Thesis, University of Warsaw
96. Gutowski M, Piela L (1988) Mol Phys 64:337
97. Kołos W, Wolniewicz L (1965) J Chem Phys 43:2429
98. Jankowski P, Jeziorski B (1999) J Chem Phys 111:1857
99. Baker JD, Freund DE, Hill RN, Morgan JD III (1990) Phys Rev A 41:1241
100. Adams WH (2005) Int J Quantum Chem (in press)
101. Larsen H, Halkier A, Olsen J, Jørgensen P (2000) J Chem Phys 112:1107
102. Jeziorski B, Moszyński R, Rybak S, Szalewicz K In Kaldor U (1989) (eds) Many-Body
     Methods in Quantum Chemistry, Lecture Notes in Chemistry, vol 52. Springer, New
     York, p 65
103. Moszyński R, Jeziorski B, Szalewicz K (1993) Int J Quantum Chem 45:409
104. Moszyński R, Jeziorski B, Ratkiewicz A, Rybak S (1993) J Chem Phys 99:8856
105. Moszyński R, Jeziorski B, Rybak S, Szalewicz K, Williams HL (1994) J Chem Phys
     100:5080
106. Moszyński R, Cybulski SM, Chałasiński G (1994) J Chem Phys 100:4998
107. Williams HL, Szalewicz K, Moszyński R, Jeziorski B (1995) J Chem Phys 103:4586
108. Patkowski K et al. to be published
109. Tang KT, Toennies JP (1984) J Chem Phys 80:3726
110. Stone AJ (1975) Mol Phys 29:1461
111. Stone AJ (1976) J Phys A 9:485
112. Tough RJA, Stone AJ (1977) J Phys A 10:1261
113. Stone AJ (1978) Mol Phys 36:241
114. Stone AJ, Tough RJA (1984) Chem Phys Lett 110:123
115. Wormer PES, Hettema H (1992) J Chem Phys 97:5592
116. Wormer PES, Hettema H (1992) POLCOR package, University of Nijmegen
117. Stone AJ (1981) Chem Phys Lett 83:233
118. Price SL, Stone AJ, Alderton M (1984) Mol Phys 52:987
119. Stone AJ, Alderton M (1985) Mol Phys 56:1047
120. Buckingham AD, Fowler PW, Stone AJ (1986) Int Rev Phys Chem 5:107
121. Stone AJ (1991) In: Maksic ZB, editor, Theoretical Models of Chemical Bonding,
     vol 4. Springer, New York, p 103

122. Stone AJ (1985) Mol Phys 56:1065
123. Fowler PW, Stone AJ (1987) J Phys Chem 91:509
124. Stone AJ (1989) Chem Phys Lett 155:102
125. Stone AJ (1989) Chem Phys Lett 155:111
126. Le Sueur RC, Stone AJ (1993) Mol Phys 78:1267
127. Williams GJ, Stone AJ (2003) J Chem Phys 119:4620
128. Stone AJ, Tong CS (1989) Chem Phys 137:121
129. Hättig C, Jansen G, Hess BA, JG Ángyán (1997) Mol Phys 91:145
130. Williams HL, Chabalowski CF (2001) J Phys Chem A 105:646
131. Misquitta AJ, Szalewicz K (2002) Chem Phys Lett 357:301
132. Misquitta AJ, Jeziorski B, Szalewicz K (2003) Phys Rev Lett 91:033201
133. Misquitta AJ, Szalewicz K (2005) J Chem Phys 122:214109
134. Misquitta AJ, Podeszwa R, Jeziorski B, Szalewicz K, to be published
135. Hesselmann A, Jansen G (2002) Chem Phys Lett 357:464
136. Hesselmann A, Jansen G (2002) Chem Phys Lett 362:319
137. Hesselmann A, Jansen G (2003) Chem Phys Lett 367:778
138. Zaremba E, Kohn W (1976) Phys Rev B 13:2270
139. Dmitriev Y, Peinel G (1981) Int J Quantum Chem 19:763
140. McWeeny R (1984) Croat Chem Acta 57:865
141. Bukowski R, Podeszwa R, Szalewicz K (2005) Chem Phys Lett 414:111
142. Szalewicz K, Podeszwa R, Misquitta AJ, Jeziorski B (2004) In: Simos T, Maroulis G (eds) Lecture Series on Computer and Computational Science: ICCMSE 2004, vol 1. VSP, Utrecht, p 1033
143. Podeszwa R, Szalewicz K (2005) Chem Phys Lett 412:488
144. Hesselmann A, Jansen G, Schütz M (2005) J Chem Phys 122:014103
145. Moldover MR, Trusler JPM, Edwards TJ, Mehl JB, Davis RS (1988) J Res Natl Inst Stand Technol 93:85
146. Williams HL, Korona T, Bukowski R, Jeziorski B, Szalewicz K (1996) Chem Phys Lett 262:431
147. Korona T, Williams HL, Bukowski R, Jeziorski B, Szalewicz K (1997) J Chem Phys 106:5109
148. Janzen AR, Aziz RA (1997) J Chem Phys 107:914
149. Hurly JJ, Moldover MR (2000) J Res Natl Inst Stand Technol 105:667
150. Jeziorska M, Bukowski R, Cencek W, Jaszuński M, Jeziorski B, Szalewicz K (2003) Coll Czech Chem Commun 68:463
151. Cencek W, Jeziorska M, Bukowski R, Jaszuński M, Jeziorski B, Szalewicz K (2004) J Phys Chem A 108:3211
152. Anderson JB (2004) J Chem Phys 120:9886
153. Lotrich VF, Szalewicz K (2000) J Chem Phys 112:112
154. Luther H, Grohman K, Fellmuth B (1996) Meterologia 33:341
155. Moldover MR (1998) J Res Natl Inst Stand Technol 103:167
156. Cencek W, Szalewicz K, Jeziorski B (2001) Phys Rev Lett 86:5675
157. Łach G, Jeziorski B, Szalewicz K (2004) Phys Rev Lett 92:233001
158. Patkowski K et al. to be published
159. Cencek W, Komasa J, Pachucki K, Szalewicz K (2005) Phys Rev Lett, submitted
160. Komasa J, Cencek W, Rychlewski J (1999) Chem Phys Lett 304:293
161. Pachucki K, Komasa J (2004) Phys Rev Lett 92:213001
162. Lach G et al. to be published
163. Pachucki K, Sapirstein J (2001) Phys Rev A 63:213001
164. Moszyński R, Heijmen TGA, van der Avoird A (1995) Chem Phys Lett 247:440

165. Moszyński R, Heijmen TGA, Wormer PES, van der Avoird A (1996) J Chem Phys 104:6997
166. Koch H, Hättig C, Larsen H, Olsen J, Jørgensen P, Fernandez B, Rizzo A (1999) J Chem Phys 111:10106
167. Maroulis G (2000) J Phys Chem A 104:4772
168. Rachet F, Chrysos M, Guillot-Noel C, Le Duff Y (2000) Phys Rev Lett 84:2120
169. Rachet F, Le Duff Y, Guillot-Noel C, Chrysos M (2000) Phys Rev A 61:062501
170. Guillot-Noel C, LeDuff Y, Rachet F, Chrysos M (2002) Phys Rev A 66:012505
171. Aziz RA (1993) J Chem Phys 99:4518
172. Boyes SJ (1994) Chem Phys Lett 221:467
173. Phelps AV, Greene CH, Burke JP Jr (2000) J Phys B 33:2965
174. Patkowski K, Murdachaew G, Fou CM, Szalewicz K (2005) Mol Phys 103:2031
175. Slavíček P, Kalus R, Paška P, Odvárková I, Hobza P, Malijevský A (2003) J Chem Phys 119:2102
176. Herman PR, LaRocque PE, Stoicheff BP (1988) J Chem Phys 89:4535
177. Murdachaew G, Szalewicz K, Jiang H, Bacic Z (2004) J Chem Phys 121:11839
178. Willey DR, Choong VE, De Lucia FC (1992) J Chem Phys 96:898
179. Zhang Y, Shi HY (2002) J Mol Struct (Theochem) 589:89
180. Lovejoy CM, Nesbitt DJ (1990) J Chem Phys 93:5387
181. Jiang H, Sarsa A, Murdachaew G, Szalewicz K, Bacic Z (2005) J Chem Phys, submitted
182. Sarsa A, Bacic Z, Moskowitz JW, Schmidt KE (2002) Phys Rev Lett 88:123401
183. Chang BT, Akin-Ojo O, Bukowski R, Szalewicz K (2003) J Chem Phys 119:11654
184. Zhu YZ Xie DQ (2004) J Chem Phys 120:8575
185. Song XG, Xu YJ, Roy PN, Jager W (2004) J Chem Phys 121:12308
186. Nauta K, Miller RE (2001) J Chem Phys 115:10254
187. Xu YJ, Jager W, Tang J, Mc Kellar ARW (2003) Phys Rev Lett 91:163401
188. Paesani F, Whaley KB (2004) J Chem Phys 121:5293
189. Callegari C, Conjusteau A, Reinhard I, Lehmann KK, Scoles G (2000) J Chem Phys 113:10535
190. Merritt JM, Douberly GE, Miller RE (2004) J Chem Phys 121:1309
191. Akin-Ojo O, Bukowski R, Szalewicz K (2003) J Chem Phys 119:8379
192. Jankowski P, Szalewicz K (1998) J Chem Phys 108:3554
193. Mc Kellar ARW (1990) J Chem Phys 93:18
194. Mc Kellar ARW (1991) Chem Phys Lett 186:58
195. Mc Kellar ARW (1998) J Chem Phys 108:1811
196. Mc Kellar ARW (2000) J Chem Phys 112:9282
197. Jankowski P, Szalewicz K (2005) J Chem Phys 123:104301
198. Sloan ED Jr (1998) Clathrate Hydrates of Natural Gases, 2nd edn. Marcel Dekker, New York
199. Akin-Ojo O, Szalewicz K (2005) J Chem Phys, in press
200. Szalewicz K, Hydrogen bond (2002) In: Meyers R et al. (eds), Encyclopedia of Physical Science and Technology, third edition, vol 7. Academic Press, San Diego, CA, p 505–538
201. Bukowski R, Sadlej J, Jeziorski B, Jankowski P, Szalewicz K, Kucharski SA, Williams HL, Rice BS (1999) J Chem Phys 110:3785
202. Millot C, Stone AJ (1992) Mol Phys 77:439
203. Millot C, Soetens JC, Costa MTCM, Hodges MP, Stone AJ (1998) J Phys Chem 102:754
204. Groenenboom GC, Wormer PES, van der Avoird A, Mas EM, Bukowski R, Szalewicz K (2000) J Chem Phys 113:6702

205. Smit MJ, Groenenboom GC, Wormer PES, van der Avoird A, Bukowski R, Szalewicz K (2001) J Phys Chem A 105:6212
206. Murdachaew G et al. to be published
207. Fermi E, Amaldi G (1934) Mem Accad Italia 6:117
208. Perdew JP, Burke K, Ernzerhof M (1996) Phys Rev Lett 77:3865
209. Adamo C, Barone V (1999) J Chem Phys 110:6158
210. Becke AD (1997) J Chem Phys 107:8554
211. Wilson PJ, Bradley TJ, Tozer DJ (2001) J Chem Phys 115:9233
212. Korona T, Moszyński R, Jeziorski B (2002) Mol Phys 100:1723
213. Hesselmann A, Jansen G (2003) Phys Chem Chem Phys 5:5010
214. Brunsteiner M, Price SL (2001) Cryst Growth Des 1:447
215. Gavezotti A (2002) Mod Simul Mat Sci Eng 10:R1
216. Price SL (2004) Cryst Eng Comm 6:344
217. Motherwell WDS, Ammon HL, Dunitz JD, Dzyabchenko A, Erk P, Gavezzotti A, Hofmann DWM, Leusen FJJ, Lommerse JPM, Mooij WTM, Price SL, Scheraga H, Schweizer B, Schmidt MU, van Eijck BP, Verwer P, Williams DE (2002) Acta Cryst B 58:647
218. Byrd EFC, Scuseria GE, Chabalowski CF (2004) J Phys Chem B 108:13100
219. Tsuzuki S, Lüthi HP (2001) J Chem Phys 114:3949
220. Tsuzuki S, Honda K, Mikami M, Tanabe K (2002) J Am Chem Soc 124:104
221. Sinnokrot MO, Valeev EF, Sherrill CD (2002) J Am Chem Soc 124:10887
222. Dion M, Rydberg H, Schröder E, Langreth DC, Lundqvist BI (2004) Phys Rev Lett 92:246401
223. von Lilienfeld OA, Tavernelli I, Rothlisberger U, Sebastiani D (2004) Phys Rev Lett 93:153004
224. Bukowski R, Szalewicz K, Chabalowski CF (1999) J Phys Chem A 103:7322
225. Sponer J, Leszczynski J, Hobza P (2001) Biopolymers 61:3
226. Hobza P, Sponer J (1999) Chem Rev 91:3247
227. Karplus M (2003) Biopolymers 68:350
228. Desfrancois C, Carles S, Schermann JP (2000) Chem Rev 100:3943
229. Kollman P, Caldwell JW, Ross WS, Pearlman DA, Case DA, DeBolt S, Cheatham TE III, Ferguson D, Siebel G (1998) Encyclopedia of Computational Chemistry. Wiley, Chichester, UK, p 11
230. MacKerell AD Jr, Brooks B III, Nilsson L, Roux B, Won Y, Karplus M (1998) Encyclopedia of Computational Chemistry. Wiley, Chichester, UK, p 271
231. van Gunsteren WF, Daura X, Mark AE (1998) Encyclopedia of Computational Chemistry. Wiley, Chichester, UK, p 1211
232. Volkov A, Koritsanszky T, Coppens P (2004) Chem Phys Lett 391:170
233. Hansen NK, Coppens P (1978) Acta Cryst A34:909
234. Coppens P (1997) X-ray Charge Densities and Chemical Bonding. Oxford University Press, New York
235. Koritsanszky T, Volkov A, Coppens P (2004) Chem Phys Lett 391:170
236. Volkov A, Li X, Koritsanszky T, Coppens P (2004) J Phys Chem A 108:4283
237. Spackman MA (1986) J Chem Phys 85:6579
238. Spackman MA (1986) J Chem Phys 85:6587
239. Spackman MA (1987) J Phys Chem 91:3179
240. Korona T, Moszyński R, Thibault F, Launay JM, Bussery-Honvault B, Boissoles J, Wormer PES (2001) J Chem Phys 115:3074

Struc Bond (2005) 116: 119–148
DOI 10.1007/430_009
© Springer-Verlag Berlin Heidelberg 2005
Published online: 8 November 2005

Interaction Potentials for Water from Accurate Cluster Calculations

Sotiris S. Xantheas

Chemical Sciences Division, Pacific Northwest National Laboratory,
902 Battelle Boulevard, Box 999, Richland, WA 99352, USA
sotiris.xantheas@pnl.gov

Abstract Recent advances in the area of ab initio theory combined with the development of efficient electronic structure software suites that take advantage of parallel hardware architectures, have resulted in our ability to obtain accurate energetics for medium-size (up to 30 molecules) clusters of water molecules. These advances offer a new route in the development of empirical interaction potentials for water, especially in the absence of experimental information regarding the cluster energetics. The use of systematically improvable methodological approaches, together with the understanding of the salient issues associated with the transferability of the models across different environments, allow for the development of hierarchical approaches in the description of the intermolecular interactions in water. The use of accurate models, which are transferable across dissimilar environments, in conjunction with quantum dynamical simulation protocols, can provide new insight into the origin of the anomalous behavior of water across the various temperature and pressure ranges that are pertinent to its chemical and biological.

1
Introduction

The abundance of water in nature, its function as a universal solvent and its role in many chemical and biological processes that are responsible for sustaining life on earth, is the driving force behind the need for understanding its behavior under different conditions, and in various environments. The availability of models that describe the properties of either pure water/ice or its mixtures with a variety of solutes ranging from simple chemical species to complex biological molecules and environmental interfaces is therefore crucial in order to be able to develop predictive paradigms that attempt to model solvation and reaction and transport in aqueous environments. In attempting to develop these models the question naturally arises "*is water different/more complex than other hydrogen bonded liquids*"? This proposition has been suggested based on the "anomalous" behavior of its macroscopic properties, such as the density maximum at 4 °C, the non-monotonic behavior of its compressibility with temperature, the anomalous behavior of its relaxation time below typical temperatures of the human body, the large value and non-monotonic dependence below 35 °C of the specific heat of constant pressure, and the smaller than expected value of the coefficient of thermal expansion [1–5].

This suggestion infers that *simple* models used to describe the relevant *inter-* and *intra*-molecular interactions will not suffice in order to reproduce the behavior of these properties over a wide temperature range. To this end, explicit microscopic level detailed information needs to be incorporated into the models in order to capture the appropriate physics at the molecular level. From the simple model of Bernal and Fowler [6], which was the first attempt to develop an empirical model for water back in 1933, this process has yielded ca. 50 different models to date. A recent review [7] provides a nearly complete account of this effort coupled to the milestones in the area of molecular simulations, such as the first computer simulation of liquid water by Barker and Watts [8] and Rahman and Stillinger [9], the first parametrization of a pair potential for water from *ab initio* calculations by Clementi and co-workers [11–14] and the first simulation of liquid water from first principles by Parrinello and Carr [15]. Many of the empirical pair potentials for water that are used widely even nowadays were developed in the early 1980s [16–19]. These early models were mainly parametrized in order to reproduce measured thermodynamic bulk properties, due to the fact that molecular level information for small water clusters was limited or even nonexistent at that time. Subsequent attempts [20–25] have focused on introducing self-consistent polarization as a means of explicitly accounting for the magnitude of the non-additive many-body effects via an induction scheme. Again the lack of accurate experimental or theoretical water cluster energetic information has prevented the assessment of the accuracy of those models.

Over the last decade, there have been tremendous advances in the experimental [26–36] and theoretical [37–40] studies of the structural, spectral and energetic properties of water clusters, as well as refinements of the macroscopic structural experimental data for liquid water [41–43]. These developments have created a unique opportunity for incorporating molecular level information into empirical potentials for water as well as using the available database for the assessment of their accuracy. The process of developing interaction potentials is by no means straightforward and/or unique, and the different approaches that are used to model the various components of the underlying physical interactions can have their own merits and shortcomings. In the subsequent sections we will outline a path for incorporating molecular level information into empirical models for water, based on quantifying the relevant interactions at the molecular level from cluster results, but having as a target the applicability of the model for macroscopic simulations.

2
Outstanding Issues

There clearly exists a need for accurate models describing the intermolecular interaction between water molecules. Furthermore, it would be highly desirable to develop hierarchical approaches in modeling those interactions, which can be systematically improvable. The incorporation of molecular level information into the models will enhance our understanding of the interplay between molecular properties and macroscopic observables. This understanding will ensure transferability across different environments such as clusters, interfaces, bulk water and ice, and will assist in expanding the current state-of-the-art in the area of force field development.

Some of the outstanding issues that are associated with the development of models based on molecular level information are the shortage or scarcity of experimental data for clusters. To this end, an alternative approach based on the use of *ab initio* results for water clusters can be adopted. Clusters offer the advantage of probing the relevant interactions at the molecular level. They moreover render a quantitative picture of the nature and magnitude of the various components of the intermolecular interactions such as exchange, dispersion and induction. They can finally serve as a vehicle for the study of the convergence of properties with cluster size. Assuming that the route of parametrizing an interaction potential from the cluster results is adopted, the question naturally arises of the level of accuracy that needs to be attained so the models can produce meaningful and accurate results for the macroscopic properties of extended systems such as the bulk liquid and ice. These issues, together with the functional form of the empirical potential function, will be considered in the subsequent sections.

3
Water Cluster Energetics

3.1
The Many-Body Expansion of the Interaction Energy

The interaction energy of a system of n molecules can be cast [44] as:

$$\Delta E_n = E(1234...n) - nE_w \tag{1}$$

$$\equiv \sum_{i=1}^{n} E(i) - nE_w \qquad \text{"Relaxation"}$$

$$+ \sum_{i=1}^{n-1} \sum_{j>i}^{n} \Delta^2 E(ij) \qquad \text{"2-body"}$$

$$+ \sum_{i=1}^{n-2} \sum_{j>i}^{n-1} \sum_{k>j}^{n} \Delta^3 E(ijk) \qquad \text{"3-body"}$$

$$+ \sum_{i=1}^{n-3} \sum_{j>i}^{n-3} \sum_{k>j}^{n-1} \sum_{l>k}^{n} \Delta^4 E(ijkl) \qquad \text{"4-body"}$$

$$+ ...$$

$$+ \Delta^n E(1234...n) \qquad \text{"n-body"}$$

where $E(i)$, $E(ij)$, $E(ijk)$, ... are the energies of the various monomers, dimers, trimers, etc. in the cluster and E_w is the energy of the isolated water molecule. The 1-body term, often referred to as the "relaxation" or "deformation" energy, represents the energy penalty for distorting the individual fragments in the equilibrium cluster geometries with respect to those in isolation. This term can be quite small, especially for systems interacting via weak interactions, and hence it has sometimes been neglected. It can, however, be as large as 5 kcal/mol for negative ion-water dimers (such as F^-H_2O and OH^-H_2O) [45, 46]. Irrespective of its magnitude, it describes a physical effect that is associated with a measurable experimental observable, namely the red shift in the infrared (IR) bands that correspond to the hydrogen bonded intramolecular stretches from their isolated gas phase monomer values. For example, in the water dimer this energy term amounts to about 0.04 kcal/mol, resulting from the elongation [38] of the hydrogen bonded OH bond lengths by around 0.07 Å and the increase of the HOH angles by 0.4° from the corresponding gas phase monomer values. These geometrical changes result in an experimental red shift of 155 cm^{-1} for the hydrogen bonded OH stretching frequency [47]. The deformation energy per individual water molecule is larger for bulk liquid water and ice as a result of the larger changes in water's internal geometry (0.972/1.008 Å for the bond lengths and 106.6/109.5° for the bond angles for liquid water [48]/ice [49], respectively) from the experimental gas phase monomer values [50] of 0.957 Å and 104.52°, a fact that in turn induces larger red shifts for the OH stretches [51].

The pairwise-additive (2-body) interaction is:

$$\Delta^2 E(ij) = E(ij) - \{E(i) + E(j)\},\qquad(2)$$

whereas the higher non-additive (3-, 4-body and larger) components are defined as:

$$\Delta^3 E(ijk) = E(ijk) - \{E(i) + E(j) + E(k)\}\qquad(3)$$
$$- \{\Delta^2 E(ij) + \Delta^2 E(ik) + \Delta^2 E(jk)\}$$
$$\Delta^4 E(ijkl) = E(ijkl) - \{E(i) + E(j) + E(k) + E(l)\}\qquad(4)$$
$$- \{\Delta^2 E(ij) + \Delta^2 E(ik)$$
$$+ \Delta^2 E(il) + \Delta^2 E(jk) + \Delta^2 E(jl) + \Delta^2 E(kl)\}$$
$$- \{\Delta^3 E(ijk) + \Delta^3 E(ijl) + \Delta^3 E(ikl) + \Delta^3 E(jkl)\}$$

etc.

The above analysis can be used to quantify the magnitude of the pairwise-additive and non-additive (cooperative) effects in molecular systems.

3.2
Magnitude of Cooperative Effects for the Water Cluster Minima ($n = 2$–6)

The application of the energy decomposition scheme at the global and local minimum geometries of the first few water clusters ($n = 3$–6) allows for the investigation of the variation of the cooperative effects with respect to the underlying hydrogen bonding network. This is because the global and local minima of the first few water clusters are associated with different connectivities, which arise from the role of the individual water molecules as hydrogen-bond donors (d), acceptors (a) and their combinations, namely double donors (dd), double acceptors (aa) or donor-acceptors (da), as graphically depicted in Fig. 1.

The geometries and relative energies of the global and higher lying local minima of the $n = 3$–6 water clusters [52] are shown in Fig. 2. In this figure, the connectivity between water molecules for each cluster geometry is indicated and the magnitude of the 3-body term is listed in parentheses. The results of the many-body decomposition are summarized in Table 1, where the percentage contribution of the 2-, 3- and 4-body terms to the cluster binding energies is listed. As can be seen from this Table, the 2-body (pairwise-additive) term is the dominant one, contributing 70–80% to the corresponding binding energies of the first few water clusters. The next most important contribution comes from the 3-body term, which amounts to 20–30% for the various isomers, while the 4-body term is quite small ($< 4\%$) and higher order

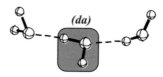

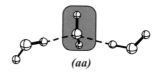

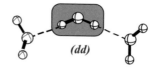

Fig. 1 Definition of the local hydrogen bonding network for a water molecule in terms of a proton donor (d) or a proton acceptor (a) with respect to neighbors

Table 1 Percentage contribution of the 2-, 3- and 4-body terms to the binding energies in the various networks of the $(H_2O)_n$, $n = 3$–6 clusters. Repulsive contributions are indicated by (R) and are responsible, together with the relaxation terms, for producing two-body components that exceed 100%

Cluster	Network	2-body	3-body	4-body
Trimer	(da,da,da)	85.1	17.7	
	(a,dd,a)	106.7	5.6 (R)	
	(d,aa,d)	104.6	3.9 (R)	
Tetramer	(da,da,da,da)	76.3	25.6	2.2
	Cage	87.3	16.5	0.6 (R)
	(aa,dd,aa,dd)	109.2	9.6 (R)	0.7
	(aa,da,dd,da)	106.8	4.5 (R)	
Pentamer	(da,da,da,da,da)	71.8	28.6	3.7
	Cage	80.4	23.1	1.3
Hexamer	Prism	80.3	22.8	1.5
	Cage	79.9	23.2	1.2
	Book	75.6	26.1	2.5
	S_6 (da,da,da,da,da,da)	69.5	29.7	4.4

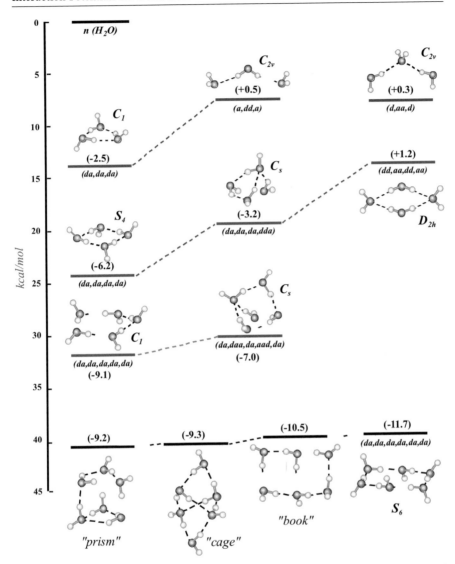

Fig. 2 Geometries and relative energies of the various water cluster global and local minima $(H_2O)_n$, $n = 3$–6. The hydrogen bonding network is denoted as (d) = donor, (a) = acceptor to neighbors and their combinations. The magnitude of the 3-body term (in kcal/mol) is listed in parentheses

terms are insignificant ($< 0.5\%$). This analysis suggests that purely 2-body interaction potentials, which are derived from the water dimer potential energy surface (such as the MCY potential), can result in errors exceeding 20% for the energetics of small clusters. In contrast, empirical potentials that have explicit 2- plus 3-body terms parameterized from the water dimer and trimer potential energy surfaces (PESs) will be more accurate, being in error by just 4% or less for the minimum energy configurations of these clusters. Extended parts of the corresponding potential energy surfaces away from the minimum energy configuration are needed, as the magnitude of the 3-body term depends on both the cluster geometry [53] as well as the connectivity of the hydrogen bonding network [40]. As an example, the 3-body term is attractive amounting to -2.5 kcal/mol (or 17.7% of the total cluster binding energy) at the global minimum ring geometry of the water trimer (all molecules exhibit donor-acceptor (*da*) connectivity). In contrast, for the higher lying "open" trimer minima, which exhibit (*a,dd,a*) and (*d,aa,d*) connectivities, (the central water molecule is a double-donor or a double-acceptor, respectively), the 3-body term is repulsive, amounting to $+0.5$ and $+0.3$ kcal/mol (or 5.6% and 3.9% of the total cluster binding energies), respectively.

The above analysis further justifies the view adopted by Clementi and coworkers [10–14] over 30 years ago, that the water dimer PES holds the key to developing a fundamental understanding of the intermolecular interactions between water molecules. The question, which remains to be addressed, is the required accuracy in the 2-body interaction in order to produce meaningful models for extended aqueous environments.

3.3
Systematic Convergence of Cluster Energetics from Electronic Structure Calculations

The cluster energies are obtained from the solution of the non-relativistic Schrödinger equation for each system. The expansion of the trial many-electron wavefunction delineates the *level of theory* (description of electron correlation), whereas the description of the constituent one-electron orbitals is associated with the choice of the *orbital basis set*. A recent review [54] outlines a path, which is based on hierarchical approaches in this double expansion in order to ensure convergence of both the correlation and basis set problems. It also describes the application of these hierarchical approaches to various chemical systems that are associated with very diverse bonding characteristics, such as covalent bonds, hydrogen bonds and weakly bound clusters.

As regards the description of the electron correlation problem it has been recognized that the coupled cluster method [55–61], which includes all possible single, double, triple, etc. excitations from a reference wavefunction, represents a viable route towards obtaining accurate energetics for hydrogen bonded dimers [62,63]. Among its variations, the CCSD(T) approximation [64,65], which includes the effects of triple excitations perturbationally, represents an excellent compromise between accuracy and computational efficiency. The CCSD(T) approximation scales as N^6 for the iterative solution of the CCSD part, plus an additional single N^7 step for the perturbative estimate of the triple excitations (N being the number of functions in the basis set). This represents substantial savings over the widely used second order perturbation level [66] of theory (MP2) which formally scales as N^5.

For the basis set expansion, the *correlation-consistent* (cc-pVnZ) orbital basis sets of Dunning and co-workers [67–69], ranging from double to quintuple zeta quality ($n =$ D, T, Q, 5), offer a systematic path in approaching the complete basis set (CBS) limit. These sets were constructed by grouping together all basis functions that contribute roughly equal amounts to the correlation energy of the atomic ground states. In this approach, functions are added to the basis sets in shells. The sets approach the complete basis set (CBS) limit, for each succeeding set in the series provides an ever more accurate description of both the atomic radial and angular spaces. The extension of those sets to include additional diffuse functions for each angular momentum function present in the standard basis yields the *augmented correlation consistent* (aug-cc-pVnZ) sets. The exponents of those additional diffuse functions were optimized for the corresponding negative ions. Application of the family of these sets to a variety of chemical systems (see [54] and references therein) ranging from the very weakly bound (by around 0.01 kcal/mol) rare gas diatomics, to intermediate strength hydrogen bonded neutral (2–5 kcal/mol), singly charged ion-water (10–30 kcal/mol) clusters and single bond diatomics (50–100 kcal/mol), to very strong (> 200 kcal/mol) multiply charged metal-water clusters and multiple bond diatomics, has permitted a *heuristic* extrapolation of the computed electronic energies and energy differences to the CBS limit. Among the various approaches that have been proposed [70–78] in order to arrive at the CBS limit, we have relied on the following two:

(i) a polynomial with inverse powers of 4 and 5 (4–5 polynomial):

$$\Delta E = \Delta E_{CBS} + \gamma/(\ell_{max} + 1)^4 + \delta/(\ell_{max} + 1)^5 \qquad (5)$$

where ℓ_{max} is the value of the highest angular momentum function in the basis set and

(ii) an exponential dependence on the cardinal number of the basis set n ($n = 2, 3, 4, 5$ for the sets of double through quintuple zeta quality, respectively):

$$\Delta E = \Delta E_{CBS} + a \cdot \exp(-\beta \cdot n). \qquad (6)$$

It should be noted that in nearly every case this "extrapolation" procedure only accounts for a very small change when compared to the "best" computed quantity with the largest basis set (usually the (aug)-cc-pV5Z or, computer resources permitting, the (aug)-cc-pV6Z). This result suggests that effective convergence of the respective properties (such as structure and relative energies) has already been achieved with the largest basis sets of this family.

It has been argued [79] that the use of a finite basis set in electronic structure calculations overestimates the interaction energy for weakly bound complexes

$$\Delta E = \Delta E_{AB}^{\alpha \cup \beta}(AB) - E_A^{\alpha}(A) - E_B^{\beta}(B), \tag{7}$$

where superscripts indicate the basis set, subscripts the molecular system and parentheses the geometries for which the energies are obtained. For instance, $\Delta E = \Delta E_{CBS} + a \cdot \exp(-\beta \cdot n)$ is the energy of complex (AB) at the complex geometry AB with the complex basis set $\cup$, etc. The problem arises because of the fact that basis functions centered on one fragment help lower the energy of a neighboring fragment and *vice versa*, a situation first termed "basis set superposition error" (BSSE) by Liu and McLean [80]. Boys and Bernardi [81] have already proposed two approaches to circumvent this problem, viz. the "point counterpoise" (pCP) and "function counterpoise" (fCP) methods, and the latter is almost exclusively adopted in electronic structure calculations [82, 83]. For a cluster of two interacting moieties (extension to larger clusters is straightforward) the fCP correction is:

$$\Delta E = (f CP) = E_{AB}^{\alpha \cup \beta}(AB) - E_{AB}^{\alpha \cup \beta}(A) - E_{AB}^{\alpha \cup \beta}(B). \tag{8}$$

It is obvious that equations 7 and 8 will not converge to the same result at the CBS limit, since the reference energies of fragments A and B are computed at different geometries (complex vs. isolated). If the BSSE correction is estimated via [84]

$$\Delta E(BSSE) = E_{AB}^{\alpha \cup \beta}(AB) - E_{AB}^{\alpha \cup \beta}(A) E_{AB}^{\alpha \cup \beta}(B) + E_{rel}^{\alpha}(A) + E_{rel}^{\beta}(B) \tag{9}$$

where

$$E_{rel}^{\alpha}(A) = E_{AB}^{\alpha}(A) - E_A^{\alpha}(A) \tag{9a}$$

$$E_{rel}^{\beta}(B) = E_{AB}^{\beta}(B) - E_B^{\beta}(B) \tag{9b}$$

represent the energy penalty for distorting the fragments from their isolated geometries to the ones that adopt in the complex, then after substitution of equations (9a) and (9b) into (9) and collecting terms the BSSE correction can be cast as

$$\Delta E(BSSE) = \Delta E - \{E_{AB}^{\alpha \cup \beta}(A) - E_{AB}^{\alpha}(A)\} - \{E_{AB}^{\alpha \cup \beta}(B) - E_{AB}^{\beta}(B)\}. \tag{10}$$

Equations 7 and 10 converge to the same result at the CBS limit, since the terms in brackets in Eq. 10 will numerically approach zero as the basis sets α and β tend towards CBS, viz.

$$\lim_{\alpha, \beta \to CBS} \Delta E(BSSE) = \lim_{\alpha, \beta \to CBS} \Delta E. \tag{11}$$

The importance of this correction, although previously recognized [85–87], has rarely been applied [88–92] until the problems arising from omitting it, especially for (i) strongly bound hydrogen bonded complexes exhibiting large fragment relaxations and (ii) calculations employing large basis sets, was pointed out [84].

4
The Water Dimer Potential Energy Surface

As noted earlier (Sect. 3.2) the water dimer PES holds the key to developing accurate descriptions of the interactions between water molecules. This fact was recognized early on by Clementi and co-workers [10–14], who almost 30 years ago attempted to parametrize the 2-body water-water interaction from Hartree-Fock (HF) and Configuration Interaction (CI) calculations of the water dimer potential energy surface (PES). The full dimer PES is a surface in 12 dimensions (6 for rigid monomers) and its full characterization will require between 10^6 (rigid monomers) or 10^{12} (flexible monomers) points, assuming a sampling of just 10 points per degree of freedom. The issue of accuracy for each point on the dimer PES can be readily seen from Figs. 3 and 4, which show the variation of the water dimer equilibrium O-O separation [38, 84, 93], R_e(O-O), and binding energy [38, 84, 94], ΔE_e, with the level of theory [MP2, MP4 and CCSD(T)] and basis set size (aug-cc-pVnZ, n = D, T, Q, 5). Results for both the BSSE-uncorrected PES (filled symbols) as well as the BSSE-corrected PES (open symbols) are shown. The optimizations were

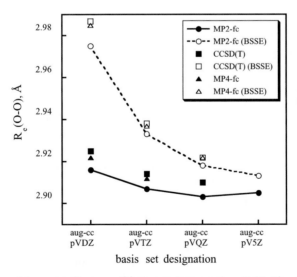

Fig. 3 Variation of the water dimer equilibrium O-O separation, R_e(O-O), with the level of theory and basis set, with (*open symbols*) and without (*filled symbols*) BSSE corrections

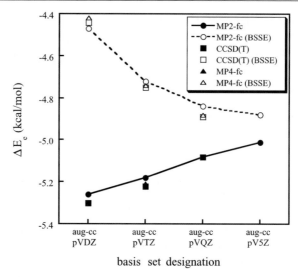

Fig. 4 Variation of the water dimer binding energy, ΔE_e, with the level of theory and basis set, with (*open symbols*) and without (*filled symbols*) BSSE corrections

performed using analytical gradients at the MP2 level and numerically at the MP4 and CCSD(T) levels of theory. The optimal geometries on the BSSE-corrected PES were obtained by numerical optimization [84] on the BSSE surface.

An alternative approach is to focus on specific cuts on the multidimensional water dimer PES, along pathways for forming hydrogen bonds between the two water molecules. These minimum energy paths (MEPs) for the approach of two water molecules forming 1 (along C_s symmetry) and 2 hydrogen bonds (along C_{2v} and C_i symmetries) as a function of the intermolecular O-O separation are shown in Fig. 5 at the MP2/aug-cc-pVTZ level of theory [95]. We note the energetic stabilization of the "doubly hydrogen bonded" configurations for $R(O\text{-}O) < 2.65$ Å. This picture does not change upon inclusion of higher correlation at the CCSD(T) level of theory or correction for BSSE [95]. It should be noted that the C_i symmetry-constrained MEP for $R(O\text{-}O) < 2.65$ Å (repulsive wall) represents the lowest energy configuration for this intermolecular separation, i.e. there is no C_1 MEP below it for this O-O range. The importance of the accurate description of the repulsive wall (shorter O-O separations) is also related to the fact that the nearest-neighbor O-O separation decreases with cluster size from the dimer to the hexamer clusters [38]. For instance, the (MP2/aug-cc-pVTZ) nearest neighbor O-O separation for the water dimer is 2.907 Å, whereas for the ring water pentamer it is in the range 2.711–2.725 Å. The minimum along the C_s MEP of Fig. 5 is the global minimum of the water dimer PES, whereas the minima along the C_i and C_{2v} symmetry-constrained MEPs of Fig. 5 are first order

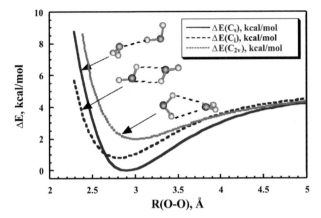

Fig. 5 Minimum energy paths (MEPs) for the approach of two water molecules along pathways forming 1 (C_s symmetry) and 2 hydrogen bonds (C_{2v} and C_i symmetries)

transition states. They represent the transition states for the "scrambling" of the hydrogen atoms in the water dimer along isomerization pathways characterized by vibration-rotational-tunneling (VRT) splittings, which have been experimentally assigned [96, 97] and theoretically modeled [98, 99]. The splittings of the energy levels and their assignments [96, 97], as well as the transition state structures and their barriers obtained at the MP2/CBS level, are depicted in Fig. 6.

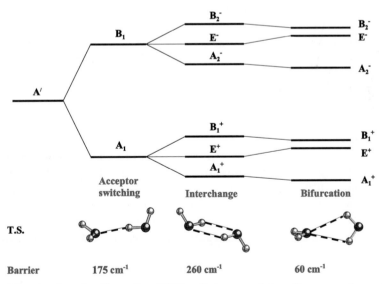

Fig. 6 Vibration-Rotation-Tunneling (VRT) splitting correlation diagram and associated (MP2/CBS) barriers for the water dimmer

The feature of the repulsive wall of the water dimer PES related to the energetic stabilization of the "doubly hydrogen bonded" configurations is not reproduced [95] by the TIP4P [16–18] and Dang-Chang (DC) [24, 25] in-

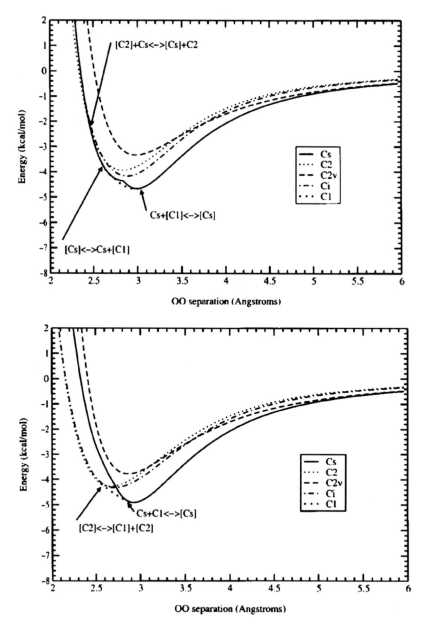

Fig. 7 Energy as a function of R(O-O) along the various MEPs for the water dimer with the ASP-W (*top*) and ASP-VRT (*bottom*) interaction potentials

teraction potentials for water. In contrast, the energetic stabilization of the doubly hydrogen bonded configurations for short O-O separations is reproduced to some extend by the MCY-CI [10–14], ASP-W [100], ASP-W4 [101], TTM [102], and TTM2-R [103] potentials. It is instructive to illustrate the effect of the experimental VRT splittings on the repulsive wall of the dimer PES. Saykally and co-workers have used their measured dimer VRT splittings [96, 97] in order to re-fit Stone's ASP-W interaction potential for water. The MEPs for the water dimer with the original ASP-W potential [100] (before the fit) are shown in Fig. 7 (top), whereas the resulting ASP-VRT potential [101] (after the fit) are shown in Fig. 7 (bottom). In this figure, bold solid lines in conjunction with the notation $[\mathfrak{I}_\alpha] \leftrightarrow [\mathfrak{I}_\beta] + \mathfrak{I}_\gamma$ are used to denote the R_c(O-O) separations where the symmetries $\mathfrak{I}_\beta$ and $\mathfrak{I}_\gamma$ (for $R > R_c$) collapse to the higher symmetry $\mathfrak{I}_\alpha$ (for $R \leq R_c$). For the cases we have considered here, this change involves exclusively the collapse of the C_2 and C_i MEPs into configurations of C_{2h} symmetry. For example, the notation $[C_{2h}] \leftrightarrow [C_i] + C_2$ indicates a collapse of the MEPs of C_i and C_2 symmetry to configurations of C_{2h} symmetry for $R \leq R_c$ Å. The square brackets are used to further denote the symmetry of the global minimum structure at that O-O separation, which was obtained by finding the minimum energy paths with no symmetry constraints. In the original ASP-W potential, the relative positions of the C_i and C_s MEPs (Fig. 7, top) do not agree with the picture suggested by the electronic structure calculations (Fig. 5). In contrast, the fitting to the experimentally measured VRT splittings has a dramatic effect on the repulsive wall of the potential, bringing the two MEPs (Fig. 7, bottom) into much closer agreement with the *ab initio* results. Saykally and co-workers mention that their fitting to the VRT splittings mostly affects the parameters of the ASP-W potential that are related to the repulsive wall of the potential. The analysis in terms of MEPs provides a graphical illustration of the effect of the fit on the features of the dimer repulsive wall.

5
Accurate Binding Energies for Water Clusters

The use of the double expansion (Sect. 3.3) in the description of the electron correlation and the orbital basis set allows for the accurate computation of water cluster binding energies. These are important data, which are currently not available experimentally, and can be used to assess the accuracy of interaction potentials for water. Table 2 shows a comparison between the MP2 and CCSD(T) binding energies obtained with the aug-cc-pVDZ and aug-cc-pVTZ basis sets for the D_{2d} isomer of the water octamer [104]. We note that the difference between the MP2 and CCSD(T) binding energies for this cluster for each basis set is < 0.1 kcal/mol. This result, together with additional calculations on medium size ($n = 3$–6) clusters [105, 106], sug-

Table 2 Binding energies (kcal/mol) of the water octamer obtained at the MP2 and CCSD(T) levels of theory with the aug-cc-pVDZ and aug-cc-pVTZ basis sets

Level of theory	Binding energy (D_e), kcal/mol	
	aug-cc-pVDZ	aug-cc-pVTZ
MP2	– 77.39	– 76.16
CCSD(T)	– 77.32	– 76.12

gests that MP2 provides an accurate description of the binding energies of small water clusters. To this end, accurate energetics for water clusters can be obtained at the MP2 level of theory, provided that the basis set expansion is carried out to the CBS limit. Table 3 lists the MP2/CBS estimates [104, 107, 108] for the binding energies of $(H_2O)_n$, $n = 2$–6, 8, 20. The variation of the cluster binding energies with basis set (aug-cc-pVDZ to aug-cc-pV5Z) for selected $(H_2O)_n$ clusters in the $n = 4$–8 range is shown in Fig. 8. The solid lines denote the BSSE-uncorrected, and the broken lines the BSSE-corrected results. The MP2/CBS limit estimates are also indicated as horizontal lines. The optimal geometries of the lowest isomers of the four low-lying families of isomers [109] for $(H_2O)_{20}$ and the variation of their MP2 binding energies [108] with basis set are shown in Fig. 9. Most of these calculations were performed using the NWChem suite of codes [110, 111] on phase 2 of the newly acquired massively parallel Hewlett-Packard su-

Table 3 MP2/CBS estimates for the binding energies (kcal/mol) of water clusters $(H_2O)_n$, $n = 2$–6, 8, 20

Water cluster $(H_2O)_n$	D_e (MP2/CBS), kcal/mol
2	– 4.98
3 cyclic	– 15.8
4 cyclic	– 27.6
5 cyclic	– 36.3
5 cage	(– 35.1)
6 prism	– 45.9
6 cage	– 45.8
6 book	– 45.6
6 cyclic	– 44.8
8 cube (D_{2d})	– 72.7 ± 0.4
8 cube (S_4)	– 72.7 ± 0.4
20, dodecahedron	– 200.1
20, fused cubes	– 212.6
20, face sharing	– 215.0
20, edge-sharing	– 217.9

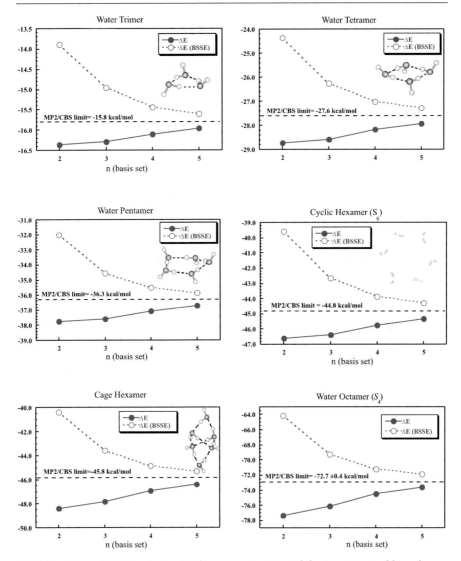

Fig. 8 Variation of $(H_2O)_n$ cluster binding energies, D_e, with basis set. *Dotted lines* denote BSSE-corrected results

percomputer at the Molecular Science Computing Facility in the William R. Wiley Environmental Molecular Sciences Laboratory at Pacific Northwest National Laboratory. The 11.4 Teraflop supercomputer consists of 1900 1.5 GHz Intel Itanium® 2 processors coupled together with Quadrics interconnect. The system has seven terabytes of memory and over 1/2 petabyte of total disk storage, making it a uniquely balanced resource for computational chemistry.

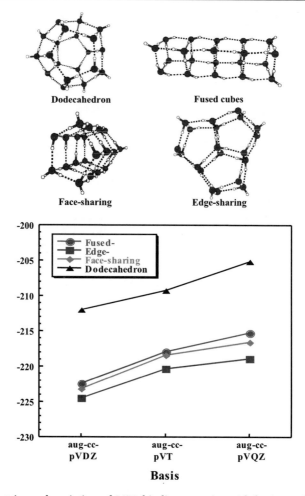

Fig. 9 Geometries and variation of MP2 binding energies with basis set for the lowest isomers within the four low-lying families of $(H_2O)_{20}$

6
Functional Form of Empirical Potentials and Appropriate Simulation Methods

6.1
Pairwise Additive vs. Many-Body Potentials

The analysis presented earlier in Sect. 3 suggests that a pairwise additive (2-body) interaction potential for water that is fitted to the water dimer PES will not be appropriate for simulations of liquid water, since it can underestimate the energetics of clusters by as much as 20%. Here we present a brief overview of some of the various choices in casting the many-body expression, $E(1, 2, 3, 4...n)$, of Eq. ?? that result in simplified descriptions of intermolecular interactions:

1. Substitute a series with just one term to describe the interactions between pairs of molecules ("effective" 2-body potential):

$$E(ijk...n) = \sum_{i<j} E_{ij}^{\text{eff}},\qquad(12)$$

where E_{ij}^{eff} is an *effective* pair-interaction energy, which is naturally different from the true pair energy, as obtained from the dimer PES. Typical representatives of this choice include the TIPnP ($n = 3, 4, 5$) [16–18, 112] and OPLS [113, 114] potentials.

2. Use an analytical expression for each term (monomer, dimer, trimer, etc.) obtained from the corresponding cluster PESs. Important issues associated with this choice are naturally, which are the important terms, and whether this series (Eq. ??) converges, and at which order. For instance, it has been previously shown [115] that the series does not converge for metals, but for hydrogen bonded systems the analysis presented in Sect. 3 suggests that it converges at $k = 3$, and that it is possible to construct an accurate (within 5%) interaction potential for water from the water dimer and trimer PES, i.e. truncating the series beyond $k = 3$. Szalewicz and co-workers [116, 117] reported a pair plus 3-body potential for water parametrized from *ab initio* calculations using 2510 dimer and 7533 trimer geometries.

3. Cast the many-body expression as

$$E(ijk...n) = \sum_i E_i + \sum_{i<j} E_{ij} + E_{\text{MB}}^{\text{eff}},\qquad(13)$$

where E_i is the 1-body (monomer) term, E_{ij} the 2-body term obtained from the dimer PES, and $E_{\text{MB}}^{\text{eff}}$ is an *effective* many-body term which:

- contains all non-additive terms to infinite order,
- is assumed to be caused exclusively by induction (additional physical terms such as 3-body exchange can be folded into other terms of the empirical form) and
- is modeled using moments (dipole) that are determined iteratively via a self-consistent scheme.

Typical candidates in this category include the polarizable TTM2-R [103] (rigid) and TTM2-F [118] (flexible) potentials, for which the terms E_i (TTM2-F) and E_{ij} (TTM2-R, TTM2-F) are fitted to the monomer and dimer PESs, and the polarizable Dang-Chang (DC) [24, 25] rigid potential, which is based on an effective two- (and many-body) term.

4. Describe all physically important terms [119], such as electrostatic, induction, dispersion, exchange repulsion, etc. with functional forms that are

parametrized from electronic structure calculations. This is the basis for the development of the family of the anisotropic site potentials (ASP) by Stone [100, 120].

6.2
The Issue of Zero-Point Energy

Most of the empirical potentials developed to simulate water have been fitted to the macroscopic structural and thermodynamic properties of the liquid. For instance, the results of classical molecular dynamics (MD) simulations are usually compared with the experimentally obtained structural radial distribution functions, bulk thermodynamic, dielectric and transport data, which are available at several temperatures and pressures. Whalley [121] has previously argued that it should be assumed that the zero-point energy should be somehow implicitly contained in the functional form/parametrization of the model in order for the results of classical MD to be directly compared to experiment. This issue has been recently discussed [95, 122] in some detail in connection with the development of rigid/flexible, pairwise-additive/many-body potentials. In effective potentials the physics (inter- and intra-molecular zero-point, cooperative effects etc.) are assumed to be *implicitly* included in the parametrization. The question still remains whether these models (which are parametrized for a specific temperature and/or phase) are appropriate for simulations outside their parametrization range, i.e. for another phase with different zero-point energy. For instance, a rigid pairwise-additive potential is an effective model for the inter-, the intra-molecular zero-point energy and the cooperative effects. Table 4 shows the classification of models into *effective* (E) and *transferable* (T) as regards the description of the inter-, the intramolecular ZPE and the cooperative effects as a function of the choice of the functional form for classical and quantum (path integral) simulation protocols.

To further illustrate the issues discussed here we consider the water cluster minimum energies obtained with the TIP4P model. These are shown in Table 5 along with the MP2/CBS limit results presented earlier in Sect. 5. The TIP4P potential is parametrized in order to reproduce the value of the enthalpy for liquid water at 300 K during classical molecular dynamics simulations. It therefore stands to reason to assume that the cluster energies obtained with the TIP4P model also correspond to enthalpies at 300 K. The seemingly "good agreement" (save for the dimer) between the TIP4P and MP2/CBS results in Table 5 is therefore deceptive, as the former (TIP4P) correspond to $\Delta H(300\,\text{K})$ values whereas the latter (MP2/CBS limit) correspond to $D_e(0\,\text{K})$. A more appropriate comparison between the two quantities is presented in Fig. 10, where the enthalpy at 300 K per molecule is plotted with cluster size for both models. The

Table 4 Classification of models into *Effective* (**E**) and *Transferable* (**T**), depending on the choice of the *intra-* and *inter-*molecular functional form and the simulation protocol used

Functional form		Simulation protocol	
Intra-	*Inter-*	Classical	Quantum (Path integral)
Rigid	Pair-wise additive	**E** (*intra*-molecular ZPE)	**E** (*intra*-molecular ZPE)
		E (*inter*-molecular ZPE)	**T** (*inter*-molecular ZPE)
		E (cooperative effects)	**E** (cooperative effects)
Rigid	Polarizable	**E** (*intra*-molecular ZPE)	**E** (*intra*-molecular ZPE)
		E (*inter*-molecular ZPE)	**T** (*inter*-molecular ZPE)
		T (cooperative effects)	**T** (cooperative effects)
Flexible	Pair-wise additive	**T** (*intra*-molecular ZPE)	**T** (*intra*-molecular ZPE)
		E (*inter*-molecular ZPE)	**T** (*inter*-molecular ZPE)
		E (cooperative effects)	**E** (cooperative effects)
Flexible	Polarizable	**T** (*intra*-molecular ZPE)	**T** (*intra*-molecular ZPE)
		E (*inter*-molecular ZPE)	**T** (*inter*-molecular ZPE)
		T (cooperative effects)	**T** (cooperative effects)

Table 5 Binding energies of the first few water clusters $(H_2O)_n$ with the TIP4P interaction potential $[\Delta H(300\,\mathrm{K})]$ and at the MP2/CBS limit $[D_e(0\,\mathrm{K})]$

n	TIP4P	MP2/CBS
2	– 6.24	– 4.98
3 cyclic	– 16.73	– 15.8
4 cyclic	– 27.87	– 27.6
5 cyclic	– 36.35	– 36.3
6 prism	– 47.27	– 45.9
6 cage	– 46.91	– 45.8
6 book	– 46.12	– 45.6
6 cyclic	– 44.38	– 44.8
8 cube (D_{2d})	– 73.02	– 72.7 ± 0.4

MP2/CBS enthalpies were obtained in the harmonic approximation using the MP2/aug-cc-pVTZ harmonic frequencies. We now note a significant difference between the two models, which arises from the inability of the effective TIP4P potential to describe the cooperative effects in small water clusters.

It is evident from the previous analysis that flexible, polarizable models in conjunction with quantum (path integral) simulations are truly transferable models, which can be used to simulate the properties of different environments that are associated with dissimilar inter- and intra-molecular zero-point energies and magnitude of cooperative effects.

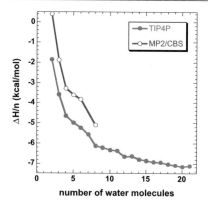

Fig. 10 Enthalpies of water clusters $(H_2O)_n$ with $n = 2-21$ for the TIP4P and TTM2-F potentials, and comparison with the results of electronic structure calculations

7
Thole-Type Interaction Potentials for Water (TTM2-R and TTM2-F)

In this section we describe the development of a flexible, polarizable interaction potential for water, which is parametrized from accurate water cluster results. It consists of the rigid (TTM2-R) [103] and flexible (TTM2-F) [95] extensions of the earlier Thole-type model (TTM) [102]. The rigid version (TTM2-R) is a 4-site model having smeared induced dipoles on the atomic sites, smeared charges of $0.574\,e$ on the hydrogen atoms and $-1.148\,e$ at a distance $d = 0.25$ Å away from the oxygen atom along the bisector of the HOH angle (M-site), as shown in Fig. 11. The OH bond lengths are 0.9572 Å and the HOH angle 104.52°, resulting to a gas phase molecular permanent dipole of 1.853 Debye. Thole's method [123] for expressing the dipole tensor in terms of the "reduced distance" $r_{ij}^{\mathrm{red}} = r_{ij}/\tilde{A}$ was used, where $\tilde{A} = (\alpha_i\alpha_j)^{1/6}$ and α_i, α_j are the polarizabilities of atoms i and j, respectively. Among the many choices proposed by Thole for the charge density we used

$$\rho(r) = \frac{1}{\tilde{A}^3}\frac{3a}{4\pi}\exp\left(-a\left(\frac{r}{\tilde{A}}\right)^3\right), \tag{14}$$

in which a is the dimensionless width parameter that is parametrized for this density ($a^{CC} = a^{CD} = 0.2$ and $a^{DD} = 0.3$). CC denotes the charge-charge,

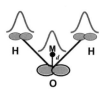

Fig. 11 The Thole-type potential for water

CD the charge-dipole and DD the dipole-dipole interactions, respectively. The only intramolecular contributions arise from the atomic dipole-dipole interactions. The induced dipoles on the O and H sites have polarizabilities of 0.837 Å^3 and 0.496 Å^3, respectively, producing a molecular polarizability of 1.433 Å^3 (experimental [124] value: 1.470 Å^3).

The total interaction for a system of N water molecules is written as

$$U_{\text{tot}} = U_{\text{pair}} + U_{\text{elec}} + U_{\text{pol}} \tag{15}$$

where

$$U_{\text{pair}} = \sum_{i=1}^{N-1} \sum_{j>i}^{N} \left(\frac{A}{r^{12}} + \frac{B}{r^{10}} + \frac{C}{r^6} \right) \tag{16}$$

$$U_{\text{elec}} = U_{CC} + U_{CD} + U_{DD} \tag{17}$$

$$U_{\text{pol}} = \sum_{i=1}^{M} \frac{\mu_i^2}{2\alpha_i} \tag{18}$$

are the pair (sums over the Oxygen sites), electrostatic and polarization components, respectively, μ_i is the induced dipole and α_i the polarizability of atomic site i, and M is the total number of induced dipole sites ($M = 3N$). The coefficients A, B and C were determined by fitting U_{tot} to symmetry-constrained energy curves [95] of the water dimer PES obtained at the MP2/aug-cc-pVTZ basis set and uniformly *scaled* to the best estimate (at the complete basis set, MP2/CBS, limit) for the minimum geometry and energy.

The flexible version of the potential (TTM2-F) is based on the coupling of the Partridge-Schwenke (PS) water monomer potential energy (PES) and dipole moment (DMS) surfaces [125] to the intermolecular part via an intramolecular charge redistribution scheme, which accounts for the change of the static molecular dipole moment due to the change of the fragment's geometry. Because the molecular dipole derivative with respect to the elongation of the OH stretch does not lie along the nuclear displacement vector [126, 127], the term "non-linear dipole moment surface" has been used. Partridge and Schwenke have produced a 245-term fit to the PES and an 84-term fit to the DMS from high level *ab initio* calculations for the water monomer. The DMS is cast in terms of geometry-dependent charges according to

$$\begin{aligned} \boldsymbol{p}^g &= q(r_{H_1O}, r_{H_2O}, \theta_{HOH}) \boldsymbol{r}_{H_1O} + q(r_{H_2O}, r_{H_1O}, \theta_{HOH}) \boldsymbol{r}_{H_2O} \\ &= q^O \boldsymbol{r}_O + q^{H_1} \boldsymbol{r}_{H_1} + q^{H_2} \boldsymbol{r}_{H_2} \end{aligned} \tag{19}$$

where $\boldsymbol{p}^g$ is the dipole moment, $\boldsymbol{r}_{H_1O} = \boldsymbol{r}_{H_1} - \boldsymbol{r}_O$, $\boldsymbol{r}_{H_2O} = \boldsymbol{r}_{H_2} - \boldsymbol{r}_O$, $q(r_{H_1O}, r_{H_2O}, \theta_{HOH}) = q^{H_1}$, and $q(r_{H_2O}, r_{H_1O}, \theta_{HOH}) = q^{H_2}$ and $q^O = -(q^{H_1} + q^{H_2})$. In other words, the partial charges *on the atomic sites* (H_1, H_2, O) are given as a function of the intramolecular geometry, viz. $q^a \equiv q^a(r_{H_1O}, r_{H_2O}, r_{H_1H_2})$.

Following the notation introduced in [95], superscripts indicate the charges on the atomic sites (H_1, H_2, O) provided by the Partridge-Schwenke DMS, whereas subscripts denote the partial charges on the charge sites (H_1, H_2, M-site) used in the model. The position along the bisector of the HOH angle of the massless "M-site" which carries the charge (instead of the oxygen atom) in the model is determined via the holonomic constraint of Reimers et al. [128–130]

$$r_M = r_O + \frac{\gamma}{2}(r_{H_1O} + r_{H_2O}), \tag{20}$$

where $\gamma = 0.4267$. In the TTM2-F potential, the partial charges q_a on the charge-bearing sites (H_1, H_2, M) are determined from the charges q^a on the nuclear sites (which are taken from the PS DMS) according to

$$q_{H_1} = q^{H_1} - \frac{\gamma}{2(1-\gamma)}q^O$$

$$q_{H_2} = q^{H_2} - \frac{\gamma}{2(1-\gamma)}q^O \tag{21}$$

$$q_M = -(q_{H_1} + q_{H_2}) = \frac{q^O}{1-\gamma}.$$

These choices satisfy the neutrality constraint, viz. $q_{H_1} + q_{H_2} + q_M = q^{H_1} + q^{H_1} + q^O = 0$ and furthermore, when substituted into Eq. 6, yield for the dipole moment $p^g = q_M r_M + q_{H_1} r_{H_1} + q_{H_2} r_{H_2} = q^O r_O + q^{H_1} r_{H_1} + q^{H_2} r_{H_2}$.

The binding energies, D_e, of the cluster minima obtained with the TTM2-R and TTM2-F interaction potentials are shown in Table 6. Also indicated are the energy differences per molecule, $\Delta E/n$, between the TTM2-F and MP2/CBS results. We note that the TTM2-F potential reproduces the water cluster energetics up to $n = 20$ with an accuracy of < 0.1 kcal/mol/molecule with respect to the MP2/CBS results, far better than any other *transferable* (flexible, polarizable) interaction potential for water for which the comparison of the minimum energy cluster energetics (D_es) to the electronic structure results is meaningful (see also the discussion in Sect. 6.2).

The effect of the accuracy of the water dimer PES used in the fitting of the potential on the accuracy of the cluster energetics for $n = 2$–21 is quantitatively seen from the results shown in Table 7. This Table lists the cluster energetics with the TTM (fitted to MP2/aug-cc-pVTZ water dimer results) and the TTM2-R potentials (fitted to uniformly *scaled* MP2/CBS water dimer results). The cluster geometries suggested by Wales and Hodges [109] have been used as starting points in the geometry optimizations. The many-body (polarization) energy is also listed in parentheses. We note that the scaling of the water dimer results to the MP2/CBS limit mainly affects the 2-body part of the corresponding cluster binding energies (the numbers in parentheses change insignificantly). The TTM2-R results are closer to the ones obtained

Table 6 Water cluster binding energies, D_e, in kcal/mol with the TTM2-R and TTM2-F potentials. $|\Delta E|/n$ denotes the absolute per molecule energy difference between the TTM2-F and MP2/CBS cluster energies: $|\Delta E|/n = |D_e(\text{TTM2-F}) - D_e(\text{MP2/CBS})|/n$

n	TTM2-R	TTM2-F	$|\Delta E|/n$
2	− 4.98	− 5.02	0.02
3 cyclic	− 15.59	− 15.90	0.03
4 cyclic	− 27.03	− 27.54	0.02
5 cyclic	− 36.05	− 36.69	0.08
5 cage	− 34.75	− 35.22	0.02
6 prism	− 45.11	− 45.86	0.01
6 cage	− 45.65	− 46.46	0.11
6 book	− 45.14	− 45.99	0.07
6 cyclic	− 44.28	− 45.03	0.04
8 cube (D_{2d})		− 73.21	0.06
8 cube (S_4)	− 71.87	− 73.24	0.06
20, dodecahedron		− 202.2	0.11
20, fused cubes		− 214.3	0.09
20, face sharing		− 214.0	0.05
20, edge-sharing		− 216.3	0.08

Table 7 Effect of the accuracy of the water dimer PES on cluster energetics

n	TTM	TTM2-R	ASP-W4
2	− 5.33 (− 0.92)	− 4.98 (− 0.91)	− 4.99
3	− 16.68 (− 6.45)	− 15.59 (− 5.14)	− 15.48
4	− 28.57 (− 9.50)	− 27.03 (− 9.57)	− 26.95
5	− 37.91 (− 13.20)	− 36.05 (− 13.36)	− 35.07
6	− 48.91 (− 14.92)	− 45.64 (− 14.89)	− 45.84
7	− 60.74 (− 18.69)	− 56.75 (− 18.63)	− 57.89
8	− 76.78 (− 23.66)	− 71.87 (− 23.65)	− 73.15
9	− 86.42 (− 27.45)	− 81.77 (− 27.73)	− 82.01
10	− 98.91 (− 31.26)	− 92.87 (− 31.32)	− 94.21
11	− 110.0 (− 34.86)	− 102.93 (− 34.82)	− 102.75
12	− 126.3 (− 39.96)	− 117.91 (− 39.75)	− 117.60
13	− 135.8 (− 42.04)	− 126.97 (− 41.91)	− 126.62
14	− 149.9 (− 47.98)	− 140.38 (− 47.92)	− 140.95
15	− 161.8 (− 51.93)	− 151.41 (− 51.86)	− 151.46
16	− 176.1 (− 56.38)	− 164.31 (− 56.03)	− 163.89
17	− 186.5 (− 58.87)	− 174.19 (− 58.60)	− 172.46
18	− 200.4 (− 64.33)	− 187.35 (− 64.10)	− 187.98
19	− 212.9 (− 67.26)	− 199.30 (− 67.18)	− 197.99
20	− 226.8 (− 73.82)	− 212.51 (− 73.72)	− 210.83
21	− 237.6 (− 74.45)	− 221.70 (− 74.14)	− 217.21

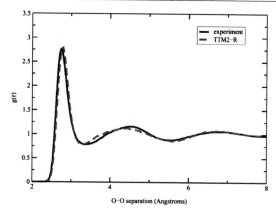

Fig. 12 Liquid water O–O radial distribution function with the TTM2-R potential

with both the ASP-W4 potential and the MP2/CBS estimates. Therefore the use of more accurate water dimer data in the fitting of the 2-body interaction yields more accurate (compared to MP2/CBS estimates) cluster energetics. This result further justifies the use of accurate electronic structure calculations in the fitting of potentials for water.

Preliminary classical molecular dynamics calculations with the effective, as regards the intra- (rigid) and inter-molecular ZPE (classical protocol), TTM2-R potential and periodic boundary conditions yield [103] liquid radial distribution functions (RDFs) in good agreement with the experimentally measured ones [131], as seen from Fig. 12. They further produce a value of 2.21×10^{-5} cm^2/s for the diffusion coefficient, which compares favorably with the experimental value [132] of 2.3×10^{-5} cm^2/s. As noted earlier, the TTM2-R is an effective model, and a full account of the magnitude of quantum effects as well as the issue of transferability between different environments (such as liquid water and the various ice forms) can be addressed only by using a truly transferable model, such as the TTM2-F in conjunction with quantum (path integral) dynamical simulation protocols.

8
Outlook

The combined simultaneous growth over the last 10 years in the area of *ab initio* theory, and the development of efficient electronic structure software suites that take advantage of parallel hardware architectures, have significantly advanced our ability to obtain accurate energetics for medium size (up to 30 molecules) clusters of water molecules. This advance provides a new route in the development of empirical interaction potentials for water, especially in the absence of experimental information regarding the cluster energetics. The use of systematically improvable methodological approaches, together with the understanding of the salient issues associated with the transferability of the models across different environments, allow for the development of hierarchical approaches in the description of the intermolecular interactions in water. The use of accurate models, which are transferable across dissimilar environments, in conjunction with quantum dynamical simulation protocols, can provide new insight into the origin of the anomalous behavior of water across the various temperature and pressure ranges that are pertinent to its chemical and biological function.

Acknowledgements This research effort has greatly benefited through valuable contributions from Drs. C. J. Burnham, G. S. Fanourgakis, E. Aprà and R. J. Harrison and helpful discussions with Drs. L. X. Dang and G. K. Schenter. This work was supported by the Division of Chemical Sciences, Geosciences and Biosciences, Office of Basic Energy Sciences, US Department of Energy. Battelle operates the Pacific Northwest National Laboratory for the Department of Energy. Calculations were performed in part using the Molecular Science Computing Facility (MSCF) in the William R. Wiley Environmental Molecular Sciences Laboratory, a national scientific user facility sponsored by the Department of Energy's Office of Biological and Environmental Research and located at Pacific Northwest National Laboratory. Additional computer resources were provided by the Office of Science, US Department of Energy.

References

1. Sciortino F, La Nave E, Scala A, Stanley HE, Starr FW (2002) Eur Phys J E 9:233
2. Stanley HE, Buldyrev SV, Canpolat M, Havlin S, Mishima O, Sadr Lahijany MR, Scala A, Starr FW (1999) Physica D, 133:453
3. Stanley HE (May 1999) MRS Bull 24(5):22–30
4. Robinson GW, Zhu SB, Singh S, Evans MW (1996) Water in Biology, Chemistry and Physics. Experimental Overviews and Computational Methodologies. World Scientific Publishing Co. Pte. Ltd, Singapore
5. Franks F (ed) (1972) Water: A comprehensive Treatise, vol 1–7. Plenum Press, New York
6. Bernal JD, Fowler RH (1933) J Chem Phys 1:515
7. Guillot B (2002) J Mol Liq 101:219
8. Barker JA, Watts RO (1969) Chem Phys Lett 3:144
9. Rahman A, Stillinger FH (1971) J Chem Phys 55:3336
10. Popkie H, Kistenmacher H, Clementi E (1973) J Chem Phys 59:1325
11. Kistenmacher H, Lie GC, Popkie H, Clementi E (1974) J Chem Phys 61:546
12. Lie GC, Clementi E (1975) J Chem Phys 62:2195
13. Matsuoka O, Clementi E, Yoshimine M (1976) J Chem Phys 64:1351
14. Lie GC, Clementi E, Yoshimine M (1976) J Chem Phys 64:2314
15. Laasonen K, Sprik M, Parrinello M, Carr R (1993) J Chem Phys 99:9080
16. Jorgensen WL (1981) J Am Chem Soc 103:335
17. Jorgensen WL (1982) J Chem Phys 77:4156
18. Jorgensen WL, Chandrasekhar J, Madura JD, Impey RW, Klein ML (1983) J Chem Phys 79:926

19. Berensen HJC, Postma JPM, van Gunsteren WF, Hermans J (1981) In: Pullman B (ed) Intermolecular Forces. Reidel, Dordrecht, p 331
20. Ahlström P, Wallqvist A, Engström S, Jönsson B (1989) Mol Phys 68:563
21. Cieplak P, Kollman P, Lybrand T (1990) J Chem Phys 92:6755
22. Sprik M (1991) J Chem Phys 95:2283
23. Kozack RE, Jordan PC (1992) J Chem Phys 96:3120
24. Dang LX (1992) J Chem Phys 97:2659
25. Dang LX, Chang TM (1997) J Chem Phys 106:8149
26. Pugliano N, Saykally RJ (1992) Science 257:1937
27. Liu K, Loeser JG, Elrod MJ, Host BC, Rzepiela JA, Pugliano N, Saykally RJ (1994) J Am Chem Soc 116:3507
28. Suzuki S, Blake GA (1994) Chem Phys Lett 229:499
29. Viant MR, Cruzan JD, Lucas DD, Brown MG, Liu K, Saykally RJ (1997) J Phys Chem A 101:9032
30. Cruzan JD, Braly LB, Liu K, Brown MG, Loeser JG, Saykally RJ (1996) Science 271:59
31. Cruzan JD, Brown MG, Liu K, Braly LB, Saykally RJ (1996) J Chem Phys 105:6634
32. Cruzan JD, Viant MR, Brown MG, Saykally RJ (1997) J Phys Chem A 101:9022
33. Liu K, Brown MG, Cruzan JD, Saykally RJ (1996) Science 271:62
34. Liu K, Brown MG, Cruzan JD, Saykally RJ (1997) J Phys Chem A 101:9011
35. Liu K, Brown MG, Carter C, Saykally RJ, Gregory JK, Clary DC (1996) Nature 381:501
36. Liu K, Brown MG, Saykally RJ (1997) J Phys Chem A 101:8995
37. Xantheas SS, Dunning TH Jr (1993) J Chem Phys 98:8037
38. Xantheas SS, Dunning TH Jr (1993) J Chem Phys 99:8774
39. Xantheas SS (1994) J Chem Phys 100:7523
40. Xantheas SS (1996) Phil Mag B 73:107
41. Soper AK (2000) Chem Phys 258:121
42. Hura G, Sorenson JM, Glaeser RM, Head-Gordon T (2000) J Chem Phys 113:9140
43. Sorenson JM, Hura G, Glaeser RM, Head-Gordon T (2000) J Chem Phys 113:9149
44. Hankins D, Moskowitz JW, Stillinger FH (1970) J Chem Phys 53:4544
45. Xantheas SS (1994) J Phys Chem 98:13489
46. Xantheas SS (1995) J Am Chem Soc 117:10373
47. Burnham CJ, Xantheas SS, Miller MA, Applegate BE, Miller RE (2002) J Chem Phys 117:1109
48. Thiessen WE, Narten AH (1982) J Chem Phys 77:2656
49. Kuhs WF, Lehman MS (1983) J Phys Chem 87:4312
50. Benedict WS, Gailar N, Plyler EK (1956) J Chem Phys 24:1139
51. Buch V, Devlin JP (1999) J Chem Phys 110:3437
52. Xantheas SS (2000) Chem Phys 258:225
53. Chalasinski G, Szczesniak MM, Cieplak C, Scheiner S (1991) J Chem Phys 94:2873
54. Dunning TH Jr (2000) J Phys Chem A 104:9062
55. Coseter F (1958) Nucl Phys 7:421
56. Coseter F, Kümmel H (1960) Nucl Phys 17:477
57. Cizek J (1966) J Chem Phys 45:4256
58. Cizek J (1969) Adv Chem Phys 14:35
59. Bartlett RJ, Purvis GD (1978) Int J Quant Chem 14:561
60. Purvis GD, Bartlett RJ (1982) J Chem Phys 76:1910
61. Kucharski SA, Bartlett RJ (1992) J Chem Phys 97:4282 and references therein
62. Peterson KA, Dunning TH Jr (1995) J Chem Phys 102:2032
63. Halkier A, Klopper W, Helgaker T, Jorgensen P, Taylor PR (1999) J Chem Phys 111:9157

64. Raghavachari K, Trucks GW, Pople JA, Head-Gordon M (1989) Chem Phys Lett 157:479
65. Raghavachari K, Pople JA, Replogle ES, Head-Gordon M (1990) J Phys Chem 94:5579
66. Møller C, Plesset MS (1934) Phys Rev 46:618
67. Dunning TH Jr (1989) J Chem Phys 90:1007
68. Kendall RA, Dunning TH Jr, Harrison RJ (1992) J Chem Phys 96:6796
69. Dunning TH Jr, Peterson KA, Woon DE (1998) In: Schleyer PvR (ed) Encyclopedia of Computational Chemistry. Wiley, New York, p 88
70. Bunge CF (1970) Theor Chim Acta 16:126
71. Termath V, Klopper W, Kutzelnigg W (1991) Chem J Phys 94:2002
72. Feller D (1992) J Chem Phys 96:6104
73. Xantheas SS, Dunning TH Jr (1993) J Phys Chem 97:18
74. Klopper W (1995) J Chem Phys 102:6168
75. Martin JML (1996) Chem Phys Lett 259:669
76. Wilson AK, Dunning TH Jr (1997) J Chem Phys 106:8718
77. Halkier A, Klopper W, Helgaker T, Jørgensen P, Taylor PR (1999) J Chem Phys 111:9157 and references therein
78. Fast PL, Sanchez L, Truhlar DG (1999) J Chem Phys 111:2921
79. Clementi E (1967) J Chem Phys 46:3851
80. Liu B, Mc Lean AD (1973) J Chem Phys 59:4557
81. Boys SF, Bernardi F (1970) Mol Phys 19:553
82. Chalasinski G, Szczesniak M (1994) Chem Rev 94:1723 and references therein
83. van Duijneveldt FB, van Duijneveldt-van de Rijdt JGCM, van Lenthe JH (1994) Chem Rev 94:1873 and references therein
84. Xantheas SS (1996) J Chem Phys 104:8821
85. Emsley J, Hoyte OPA, Overill RE (1978) J Am Chem Soc 100:3303
86. Smit PH, Derissen JL, van Duijneveldt FB (1978) J Chem Phys 69:4241
87. van Lenthe JH, van Duijneveldt-van de Rijdt JGCM, van Duijneveldt FB (1987) Adv Chem Phys 69:521
88. Leclercq JM, Allavena M, Bouteiller Y (1983) J Chem Phys 78:4606
89. Kendall RA, Simons J, Gutowski M, Chalasinski G (1989) J Phys Chem 93:621
90. Eggenberger R, Gerber S, Huber H, Searles D (1991) Chem Phys Lett 183:223
91. Mayer I, Surjan PR (1992) Chem Phys Lett 191:497
92. van Duijneveldt-van de Rijdt JGCM, van Duijneveldt FB (1992) J Chem Phys 97:5019
93. Xantheas SS, Dunning TH Jr (1998) Ab-initio Characterization of Water and Negative Ion-Water Clusters. In: Bacic Z, Bowman JM (eds) Advances in Molecular Vibrations and Collision Dynamics, vol 3. JAI Press, Stanford, Conneticut, p 281–309
94. Feyereisen MW, Feller D, Dixon DA (1996) J Phys Chem 100:2993–2997
95. Burnham CJ, Xantheas SS (2002) J Chem Phys 116:1479
96. Keutsch FN, Goldman N, Karyakin EN, Harker HA, Sanz ME, Leforestier C, Saykally RJ (2001) Faraday Discuss 118:79
97. Goldman N, Fellers RS, Brown MG, Braly LB, Keoshian CJ, Leforestier C, Saykally RJ (2002) J Chem Phys 116:10148 and references therein
98. Watanabe Y, Taketsugu T, Wales DJ (2004) J Chem Phys 120:5993
99. Wales DJ (1999) In: Jellinek J (ed) Theory of Atomic and Molecular Clusters. Springer-Verlag, Heidelberg, pp 86–110
100. Millot C, Stone A (1992) Mol Phys 77:439
101. Fellers RS, Leforestier C, Braly LB, Brown MG, Saykally RJ (1999) Science 284:945
102. Burnham CJ, Li JC, Xantheas SS, Leslie M (1999) J Chem Phys 110:4566
103. Burnham CJ, Xantheas SS (2002) J Chem Phys 116:1500

104. Xantheas SS, Aprà E (2004) J Chem Phys 120:823
105. Nielsen IMB, Seidl ET, Janssen CL (1999) J Chem Phys 110:9435
106. Xantheas SS, to be published
107. Xantheas SS, Burnham CJ, Harrison RJ (2002) J Chem Phys 116:1493
108. Fanourgakis GS, Aprà E, Xantheas SS (2004) J Chem Phys 121:2655
109. Wales DJ, Hodges MP (1998) Chem Phys Lett 286:65
110. Kendall RA, Aprà E, Bernholdt DE, Bylaska EJ, Dupuis M, Fann GI, Harrison RJ, Ju J, Nichols JA, Nieplocha J, Straatsma TP, Windus TL, Wong AT (2000) Comp Phys Comm 128:260
111. High Performance Computational Chemistry Group (2003) NWChem, A Computational Chemistry Package for Parallel Computers, Version 4.6. Pacific Northwest National Laboratory, Richland, WA 99352, USA
112. Mahoney MW, Jorgensen WL (2001) J Chem Phys 115:10758
113. Jorgensen WL, Tirado-Rives J (1988) J Am Chem Soc 110:1657–1666
114. Damm W, Frontera A, Tirado-Rives J, Jorgensen WL (1997) J Comp Chem 18:1955
115. Kaplan IG, Hernandez-Cobos J, Ortega-Blake II, Novaro O (1996) Phys Rev A 53:2493–2500
116. Groenenboom GC, Mas EM, Bukowski R, Szalewicz K, Wormer PES, van der Avoird A (2000) Phys Rev Lett 84:4072
117. Mas EM, Bukowski R, Szalewicz K (2003) J Chem Phys 118:4404
118. Burnham CJ, Xantheas SS (2002) J Chem Phys 116:5115
119. Stone AJ (1997) Theory of Intermolecular Forces. Clarendon Press, Oxford
120. Millot C, Soetens JC, Martins Costa MTC, Hodges MP, Stone AJ (1998) J Phys Chem A 102:54
121. Whalley E (1984) J Chem Phys 81:4087
122. Burnham CJ, Xantheas SS (2004) J Mol Liq 110:177
123. Thole BT (1981) Chem Phys 59:341
124. Murphy WF (1977) J Chem Phys 67:5877
125. Partridge H, Schwenke DW (1997) J Chem Phys 106:4618
126. Ikawa SI, Maeda S (1968) Spectrochim Acta, Part A 24:655
127. Whalley E, Klug DD (1986) J Chem Phys 84:78
128. Reimers JR, Watts RO, Klein ML (1981) Chem Phys 64:95
129. Reimers JR, Watts RO (1984) Chem Phys 85:83
130. Suhm MA, Watts RO (1991) Mol Phys 73:463
131. Soper AK (2000) Chem Phys 258:121
132. Krynicki K, Green CD, Sawyer DW (1978) Discuss Faraday Soc 66:199

Author Index Volumes 101–116

Subject Index

Printing: Krips bv, Meppel
Binding: Stürtz, Würzburg